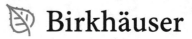

Frontiers in Mathematics

This series is designed to be a repository for up-to-date research results which have been prepared for a wider audience. Graduates and postgraduates as well as scientists will benefit from the latest developments at the research frontiers in mathematics and at the "frontiers" between mathematics and other fields like computer science, physics, biology, economics, finance, etc. All volumes are online available at SpringerLink.

More information about this series at http://www.springer.com/series/5388

Arno van den Essen • Shigeru Kuroda •
Anthony J. Crachiola

Polynomial Automorphisms and the Jacobian Conjecture

New Results from the Beginning of the 21st Century

Arno van den Essen
Department of Mathematics
Radboud University Nijmegen
Nijmegen, The Netherlands

Shigeru Kuroda
Department of Mathematical Sciences
Tokyo Metropolitan University
Hachioji-shi, Tokyo, Japan

Anthony J. Crachiola
College of Science, Engineering & Technology
Saginaw Valley State University
Saginaw, MI, USA

ISSN 1660-8046   ISSN 1660-8054 (electronic)
Frontiers in Mathematics
ISBN 978-3-030-60533-9   ISBN 978-3-030-60535-3 (eBook)
https://doi.org/10.1007/978-3-030-60535-3

Mathematics Subject Classification: 14R10, 14R15, 14L24

This book is published under the imprint Birkhäuser, www.birkhauser-science.com, by the registered company Springer Nature Switzerland AG.
The registered company address is: Gewerbestrasse 11, 6330 Cham, Switzerland

For Toshio Kuroda and Michiko Kuroda
For Jennifer, Joseph, and Renee
For Sandra and Raïssa

Preface

In March 2015, I received an email from Clemens Heine, Executive Editor for Birkhäuser, asking me if I was interested in preparing an updated new edition of my book *Polynomial Automorphisms and the Jacobian Conjecture*, which appeared in 2000. Having thought some minutes about this proposal, I realized that many new exciting results have been obtained since the publication of that book some twenty years ago. To mention a few, the solution of Nagata's conjecture by Shestakov–Umirbaev, the complete solution of Hilbert's fourteenth problem by Kuroda, the equivalence of the Jacobian Conjecture and the Dixmier Conjecture by Tsuchimoto and independently by Belov-Kanel and Kontsevich, the symmetric reduction by de Bondt and myself, the theory of Mathieu–Zhao spaces by Wenhua Zhao, and finally the counterexamples to the Cancellation Problem in positive characteristic by Neena Gupta.

In order to give a good account of all these new developments, I asked the help of two experts, both excellent expository writers, Anthony J. Crachiola and Shigeru Kuroda. I asked Tony to cover the new results related to the Cancellation Problem and Shigeru to expose his results on Hilbert's fourteenth problem and the Shestakov–Umirbaev theory. During the writing of my own contributions, it became clear that together we would have enough material to write a new book.

The contents of this book are arranged as follows. In the first chapter, written by Shigeru Kuroda, the author extends the results obtained by Shestakov and Umirbaev and gives a useful criterion to decide if a given polynomial automorphism in dimension three is tame. His exposition is completely self-contained. As an application, it is shown that Nagata's famous automorphism is indeed wild, as originally conjectured by Nagata in 1972.

The second chapter, which is also written by Kuroda, gives a complete solution of Hilbert's fourteenth problem due to the author and includes his latest results.

The third chapter, written by Anthony J. Crachiola, discusses methods used to study problems related to polynomials in positive characteristic, including the Makar-Limanov and Derksen invariants, higher derivations, gradings, etcetera. These methods are used to explain the counterexamples to the Cancellation Problem in positive characteristic due to Neena Gupta.

The last two chapters form my own contribution. In Chap. 4, it is shown that the Jacobian Conjecture is equivalent to the Dixmier Conjecture, the Poisson Conjecture, and the Unimodular Conjecture. Furthermore, a p-adic formulation of the Jacobian Conjecture is given, due to Lipton and the author. At the end of the chapter, a false "proof" of the Jacobian Conjecture is constructed. It is left to the reader to find the error!

In Chap. 5, the theory of Mathieu–Zhao spaces, mainly due to Wenhua Zhao, is developed and various conjectures, all implying the Jacobian Conjecture, are discussed: the Vanishing Conjecture, the Generalized Vanishing Conjecture, the Image Conjecture, and the Gaussian Moment Conjecture.

After the last chapter, a list of corrections to my book [117] is given. At the time that book was published, it was the only one covering a large part of the young field of polynomial automorphisms, which belongs to the larger field of Affine Algebraic Geometry. During the last 20 years, several books in the field of Affine Algebraic Geometry were published. Two of them in the series *Encyclopaedia of Mathematical Sciences*, namely *Computational Invariant Theory* by Harm Derksen and Gregor Kemper [29] and *Algebraic Theory of Locally Nilpotent Derivations* by Gene Freudenburg [48]. A second enlarged edition of this monograph appeared in 2017. In 2016, the book *The Asymptotic Variety of Polynomial Maps* was published by Ronen Peretz [99], describing his approach to the two-dimensional Jacobian Conjecture.

Also, several survey papers covering a more geometric approach to polynomial maps appeared on arXiv, such as Masayoshi Miyanishi's *Lectures on Geometry and Topology of Polynomials - Surrounding the Jacobian Conjecture* [87] and the recent paper *On the Geometry of the Automorphism Groups of Affine Varieties* by Jean-Philippe Furter and Hanspeter Kraft [52]. Finally, we like to mention the very interesting recent paper *The Jacobian Conjecture Fails for Pseudo-Planes* by Adrien Dubouloz and Karol Palka [37].

Nijmegen, The Netherlands   Arno van den Essen
July 2019

Acknowledgments

It is my great pleasure to thank all those who have directly or indirectly supported my research so far. The works in Chap. 1 or 2 of this book might not have existed without them. Special thanks go to Arno van den Essen for inviting me to contribute to this book, published as a sequel of his book [117]. I was happy to accept his invitation, because [117] is special to me. About 20 years ago, I wrote my master's thesis about initial algebras based on my own interest [67]. At that time, I already had a feeling that research in that direction would yield important results. However, it was not easy to create mathematics as I was thinking. After groping in the dark, I arrived at Arno's book [117], where I found exactly what I had been looking for. In my view, [117] is a serious attempt to understand the real nature of polynomial rings, whose value will be timeless.

I would like to acknowledge the support of JSPS KAKENHI grant number 18K03219.

Shigeru Kuroda

I am so grateful to Arno van den Essen for inviting me to be a part of this project. I especially appreciate his attitude toward writing: no deadlines, use your best judgment, and have fun! It was a lot of fun. Thanks to my home institution, Saginaw Valley State University, for providing an excellent work environment and all the support necessary to write. Special thanks to Lenny Makar-Limanov for teaching me about locally nilpotent derivations and for mentoring me in the years since. Finally, thanks to my wife Jennifer and our children Joseph and Renee. You all are the most perfect inspiration, motivation, and distraction all at once.

Anthony J. Crachiola

First of all, I want to thank Anthony J. Crachiola and Shigeru Kuroda for their beautiful contributions to this book. Without them, this "second volume" of *Polynomial Automorphisms and the Jacobian Conjecture* would never be written.

Next, I want to thank Michiel de Bondt. His many ideas have had a great influence both on my understanding of the Jacobian Conjecture and on the content of the last two chapters of this book.

Then there is of course Wenhua Zhao. His beautiful ideas shaped the last chapter and enriched our view of the Jacobian Conjecture. Furthermore, I like to thank him for bringing my wife, Sandra, in contact with the Chinese acupuncturist Dr. Zhang during the conference we organized in Tianjin in July 2014. This meeting and several later meetings with Dr. Zhang changed our lives and that is what I am the most grateful for.

Also, I like to thank Hongbo Guo for inviting me to Tianjin several times. Next, I want to thank Pascal Adjamagbo and Harm Derksen for our long friendship and fruitful mathematical cooperation. I want to thank Jan Schoone for reading several parts of this book, giving suggestions for its improvement, and helping me with the editing of the final text.

A special word of thanks to Clemens Heine from Birkhäuser for the excellent cooperation during this book project over the years.

Finally, I like to thank my wife, Sandra, and our daughter, Raissa: the joy and happiness they bring into my life every day cannot be expressed in words.

Arno van den Essen

Contents

The Shestakov–Umirbaev Theory and Nagata's Conjecture

1.1   The Shestakov–Umirbaev Theory

As discussed in the introduction of [117], Nagata [94] conjectured that there exist wild polynomial automorphisms in three variables (Conjecture 1.1.1). In 2004, after [117] was published, Shestakov–Umirbaev [109, 110] showed that the conjecture is true if the coefficient field is of characteristic zero. The purpose of this chapter is to give an introduction to the Shestakov–Umirbaev theory. We give a self-contained proof of a wildness criterion of polynomial automorphisms in three variables (Sect. 1.1.4). Using this criterion, we can easily check the wildness of Nagata's famous automorphism (Exercise 11).

We emphasize that the Shestakov–Umirbaev theory is not a theory of wild automorphisms, but a theory of tame automorphisms. This may sound strange. However, if one would like to say that an automorphism is wild, i.e., not tame, one first needs to know what the tame automorphisms are. Shestakov and Umirbaev obtained a condition which *every tame automorphism satisfies*. It is not difficult to check that Nagata's automorphism does not satisfy the condition.

We note that the "Shestakov–Umirbaev theory" we present here is the one modified by the author [73, 74]. The main differences from the original theory are as follows:

- One of the most important tools in the Shestakov–Umirbaev theory is a certain inequality for estimating degrees of polynomials (cf. [109], Theorem 3). The author improved this result in [73], which makes the argument more simple and precise.
- Shestakov and Umirbaev considered total degrees, while the author considered more generally weighted degrees. This generalization is of great help in applications. Also, the proof of Nagata's conjecture becomes slightly simpler by using an "independent" weight (cf. Sect. 1.1.2).

© The Author(s), under exclusive license to Springer Nature Switzerland AG 2021
A. van den Essen et al., *Polynomial Automorphisms and the Jacobian Conjecture*,
Frontiers in Mathematics, https://doi.org/10.1007/978-3-030-60535-3_1

Notation and Convention Throughout this chapter, let k be a field of characteristic zero except for Sect. 1.1.1, and let $k[\boldsymbol{x}] = k[x_1, \ldots, x_n]$ be the polynomial ring in n variables over k. We often use the letters F and G to denote the elements of $k[\boldsymbol{x}]^r$ for $r \geq 1$, and write $F = (f_1, \ldots, f_r)$ and $G = (g_1, \ldots, g_r)$ without mentioning it. We write $S_l :=$ $\{f_1, \ldots, \widehat{f_l}, \ldots, f_r\}$ for $1 \leq l \leq r$. For each permutation $\sigma \in \mathfrak{S}_r$, we define $F_\sigma :=$ $(f_{\sigma(1)}, \ldots, f_{\sigma(r)})$. We denote by $\mathscr{T}$ the set of $F \in k[\boldsymbol{x}]^3$ such that f_1, f_2, and f_3 are algebraically independent over k. We identify each $F \in k[\boldsymbol{x}]^r$ with the substitution map $k[x_1, \ldots, x_r] \to k[\boldsymbol{x}]$ defined by $x_i \mapsto f_i$ for each i. Then, for $F \in k[\boldsymbol{x}]^n$ and $G \in k[\boldsymbol{x}]^r$, the composite

$$k[x_1, \ldots, x_r] \xrightarrow{G} k[\boldsymbol{x}] \xrightarrow{F} k[\boldsymbol{x}]$$

is written as $FG = (g_1(f_1, \ldots, f_n), \ldots, g_r(f_1, \ldots, f_n))$.

1.1.1   Nagata's Conjecture

In this section, let k be any field. Denote by $\mathrm{Aut}_k\, k[\boldsymbol{x}]$ the automorphism group of the k-algebra $k[\boldsymbol{x}]$. We remark that $F \in k[\boldsymbol{x}]^n$ belongs to $\mathrm{Aut}_k\, k[\boldsymbol{x}]$ if and only if $k[f_1, \ldots, f_n] = k[\boldsymbol{x}]$, i.e., $F : k[\boldsymbol{x}] \to k[\boldsymbol{x}]$ is surjective, since $\mathrm{tr.deg}_k(k[\boldsymbol{x}]/\ker F) = n$ implies $\ker F = (0)$. For example, we have

$$(x_1, \ldots, x_n)A + (b_1, \ldots, b_n) \in \mathrm{Aut}_k\, k[\boldsymbol{x}] \tag{1.1.1}$$

for each $A \in GL(n, k)$ and $b_1, \ldots, b_n \in k$, and

$$(x_1, \ldots, x_{l-1}, x_l + f, x_{l+1}, \ldots, x_n) \in \mathrm{Aut}_k\, k[\boldsymbol{x}] \tag{1.1.2}$$

for each $1 \leq l \leq n$ and $f \in k[x_1, \ldots, \widehat{x_l}, \ldots, x_n]$. Automorphisms of $k[\boldsymbol{x}]$ as in (1.1.1) and (1.1.2) are called **affine automorphisms** and **elementary automorphisms**, respectively. The **tame subgroup** $T(n, k)$ is the subgroup of $\mathrm{Aut}_k\, k[\boldsymbol{x}]$ generated by all the affine automorphisms and elementary automorphisms. We say that $F \in \mathrm{Aut}_k\, k[\boldsymbol{x}]$ is **tame** if F belongs to $T(n, k)$, and **wild** otherwise.

Clearly, we have $\mathrm{Aut}_k\, k[x_1] = T(1, k)$. Due to Jung [63] and van der Kulk [129], $\mathrm{Aut}_k\, k[x_1, x_2] = T(2, k)$ also holds true. Nagata conjectured that $\mathrm{Aut}_k\, k[x_1, x_2, x_3] \neq T(3, k)$ and gave the following candidate example of a wild automorphism (see [94], Part 2, Conjecture 3.1).

Conjecture 1.1.1 (Nagata) $(f_1, f_2, f_3) \notin T(3, k)$, where

$$f_1 = x_1 - 2(x_1x_3 + x_2^2)x_2 - (x_1x_3 + x_2^2)^2 x_3$$

$$f_2 = x_2 + (x_1x_3 + x_2^2)x_3 \tag{1.1.3}$$

$$f_3 = x_3.$$

Exercise 1 Show that $k[f_1, f_2, f_3] = k[x_1, x_2, x_3]$ for f_1, f_2, and f_3 in (1.1.3).
[Note that $x_1x_3 + x_2^2 = f_1f_3 + f_2^2 \in k[f_1, f_2, f_3]$.]

About 30 years after Nagata's book [94] was published, Shestakov–Umirbaev [109, 110] finally settled this conjecture when k is of characteristic zero.

Theorem 1.1.2 (Shestakov–Umirbaev) *Conjecture 1.1.1 is true if k is of characteristic zero.*

At present, however, it is not known whether $\mathrm{Aut}_k k[\boldsymbol{x}] = T(n, k)$ holds when $n = 3$ and k is of positive characteristic, or when $n \geq 4$.

Exercise 2

(1) Show that $T(n, k)$ is generated by the automorphisms of the form

$$(x_1, \ldots, x_{l-1}, \alpha x_l + f, x_{l+1}, \ldots, x_n),$$

where $1 \leq l \leq n$, $\alpha \in k^*$ and $f \in k[x_1, \ldots, \widehat{x_l}, \ldots, x_n]$.
[By linear algebra, every element of $GL(n, k)$ is changed to the identity matrix by an iteration of column operations.]
(2) Let $E(n, k)$ be the subgroup of $\mathrm{Aut}_k k[\boldsymbol{x}]$ generated by all the elementary automorphisms, and let

$$D(n, k) := \{(\alpha_1 x_1, \ldots, \alpha_n x_n) \mid \alpha_1, \ldots, \alpha_n \in k^*\}.$$

Show that $T(n, k) = D(n, k)E(n, k)$.

1.1.2 Weighted Grading

Let Γ be a **totally ordered $\mathbb{Q}$-vector space**, i.e., a $\mathbb{Q}$-vector space equipped with a total ordering such that

$$\alpha \leq \beta \text{ implies } \alpha + \gamma \leq \beta + \gamma \text{ for each } \alpha, \beta, \gamma \in \Gamma. \tag{1.1.4}$$

For example, $\mathbb{R}$ is a $\mathbb{Q}$-vector space and satisfies (1.1.4) for the standard ordering. The $\mathbb{Q}$-vector space $\mathbb{Q}^n$ satisfies (1.1.4) for the lexicographic order.

We fix a **weight** $\mathbf{w} = (w_1, \ldots, w_n) \in \Gamma^n$ with $w_1, \ldots, w_n > 0$ and consider the **w-weighted grading** $k[\boldsymbol{x}] = \bigoplus_{\gamma \in \Gamma} k[\boldsymbol{x}]_\gamma$, where $k[\boldsymbol{x}]_\gamma$ is the k-vector space generated by monomials $x_1^{i_1} \cdots x_n^{i_n} \in k[\boldsymbol{x}]$ with $i_1 w_1 + \cdots + i_n w_n = \gamma$. Note that $k[\boldsymbol{x}]_\alpha k[\boldsymbol{x}]_\beta \subset k[\boldsymbol{x}]_{\alpha+\beta}$ for each $\alpha, \beta \in \Gamma$.

Exercise 3 Assume that $h \in k[\boldsymbol{x}]_\delta$, $h_1 \in k[\boldsymbol{x}]_{\delta_1}, \ldots, h_r \in k[\boldsymbol{x}]_{\delta_r}$ are nonzero, where $\delta, \delta_1, \ldots, \delta_r \in \Gamma$. Show that $h \in k[h_1, \ldots, h_r]$ implies

$$\delta \in \mathbb{Z}_{\geq 0}\delta_1 + \cdots + \mathbb{Z}_{\geq 0}\delta_r := \{a_1\delta_1 + \cdots + a_r\delta_r \mid a_1, \ldots, a_r \in \mathbb{Z}_{\geq 0}\}.$$

Now, take any $f = \sum_{\gamma \in \Gamma} f_\gamma \in k[\boldsymbol{x}]$, where $f_\gamma \in k[\boldsymbol{x}]_\gamma$. If $f \neq 0$, we define the **w-degree** of f by

$$\deg_{\mathbf{w}} f := \max\{\gamma \in \Gamma \mid f_\gamma \neq 0\}$$

and set $f^{\mathbf{w}} := f_{\deg_{\mathbf{w}} f}$. If $f = 0$, we define $\deg f := -\infty$ and $f^{\mathbf{w}} := 0$. Then, for each $f, g \in k[\boldsymbol{x}]$, we have

$$\deg_{\mathbf{w}} fg = \deg_{\mathbf{w}} f + \deg_{\mathbf{w}} g \qquad \text{and} \qquad (fg)^{\mathbf{w}} = f^{\mathbf{w}} g^{\mathbf{w}}. \qquad (1.1.5)$$

Example 1.1.3

(1) If $\mathbf{w} = (1, \ldots, 1) \in \mathbb{R}^n$, then $\deg_{\mathbf{w}} f$ is the same as the total degree of f.
(2) Let $\Gamma = \mathbb{Q}^3$ with the lexicographic order, and let $\mathbf{w} = (\mathbf{e}_1, \mathbf{e}_2, \mathbf{e}_3)$, where $\mathbf{e}_1$, $\mathbf{e}_2$, and $\mathbf{e}_3$ are the coordinate unit vectors of $\mathbb{Q}^3$. Then, for f_1, f_2, and f_3 in (1.1.3), we have $f_1^{\mathbf{w}} = -x_1^2 x_3^3$, $f_2^{\mathbf{w}} = x_1 x_3^2$, and $f_3^{\mathbf{w}} = x_3$, and

$$\deg_{\mathbf{w}} f_1 = (2, 0, 3), \quad \deg_{\mathbf{w}} f_2 = (1, 0, 2), \quad \text{and} \quad \deg_{\mathbf{w}} f_3 = (0, 0, 1).$$

Exercise 4 Show the following:

(1) $L := \{f/g \mid f, g \in k[\boldsymbol{x}]_\gamma, \ g \neq 0, \ \gamma \in \Gamma\}$ is a subfield of the rational function field $k(x_1, \ldots, x_n)$.
(2) Every element h of $k[\boldsymbol{x}] \setminus k$ is transcendental over L. [For any $\gamma \in \Gamma$, $l \geq 0$ and $c_0, \ldots, c_l \in k[\boldsymbol{x}]_\gamma$ with $c_l \neq 0$, we have $(\sum_{i=0}^{l} c_i h^i)^{\mathbf{w}} = c_l^{\mathbf{w}} (h^{\mathbf{w}})^l \neq 0$.]

We say that $\mathbf{w}$ is **independent** if $w_1, \ldots, w_n$ are linearly independent over $\mathbb{Q}$. For example, $\mathbf{w} = (1, \sqrt{2}, \sqrt{3}) \in \mathbb{R}^3$ is independent. If $\mathbf{w}$ is independent, then $\deg_{\mathbf{w}} x_1^{i_1} \cdots x_n^{i_n} = \sum_{l=1}^{n} i_l w_l$'s are different for different $(i_1, \ldots, i_n)$'s. Hence, $f^{\mathbf{w}}$ is always a monomial. Moreover, we have

$$\deg_{\mathbf{w}} f = \deg_{\mathbf{w}} g \iff f^{\mathbf{w}} \approx g^{\mathbf{w}} \qquad \text{for} \qquad f, g \in k[\mathbf{x}]. \tag{1.1.6}$$

Here, we write $f \approx g$ (resp., $f \not\approx g$) if f and g are linearly dependent (resp., linearly independent) over k.

Exercise 5 Consider the following conditions for $f_1, \ldots, f_r \in k[\mathbf{x}] \setminus \{0\}$:

(a) $\deg_{\mathbf{w}} f_1, \ldots, \deg_{\mathbf{w}} f_r$ are linearly independent over $\mathbb{Q}$.
(b) $f_1^{\mathbf{w}}, \ldots, f_r^{\mathbf{w}}$ are algebraically independent over k.

(1) Show that (a) implies (b).
   [$\deg_{\mathbf{w}}(f_1^{\mathbf{w}})^{i_1} \cdots (f_r^{\mathbf{w}})^{i_r}$'s are different for different $(i_1, \ldots, i_r)$'s.]
(2) Show that (b) implies (a) when $\mathbf{w}$ is independent.
   [If (a) is false, then $\deg_{\mathbf{w}} f_1^{i_1} \cdots f_r^{i_r} = \deg_{\mathbf{w}} f_1^{j_1} \cdots f_r^{j_r}$ for some distinct $(i_1, \ldots, i_r), (j_1, \ldots, j_r) \in (\mathbb{Z}_{\geq 0})^r$. This implies $(f_1^{i_1} \cdots f_r^{i_r})^{\mathbf{w}} \approx (f_1^{j_1} \cdots f_r^{j_r})^{\mathbf{w}}$ by (1.1.6).]

Proposition 1.1.4 *Let $f \in k[\mathbf{x}]_\alpha \setminus k$ and $g \in k[\mathbf{x}]_\beta \setminus k$, where $\alpha, \beta \in \Gamma$. If f and g are algebraically dependent over k, then $f^q \approx g^p$ holds for some $p, q \geq 1$ with $\gcd(p, q) = 1$.*

Proof Note that $f^{\mathbf{w}} = f$ and $g^{\mathbf{w}} = g$. Hence, by Exercise 5 (1), we can find $p, q \geq 1$ such that $q \deg f = p \deg g$ and $\gcd(p, q) = 1$. Then, f^q/g^p lies in the field L of Exercise 4. Hence, g is transcendental over $k(f^q/g^p)$. Since $\text{tr.deg}_k k(f^q/g^p, g) = 1$, it follows that $f^q/g^p \in k$. □

We study tuples of elements of $k[\mathbf{x}]$ by means of $\mathbf{w}$-weighted gradings. For each $F \in k[\mathbf{x}]^r$ with $r \geq 1$, we define

$$\deg_{\mathbf{w}} F := \deg_{\mathbf{w}} f_1 + \cdots + \deg_{\mathbf{w}} f_r \quad \text{and} \quad F^{\mathbf{w}} := (f_1^{\mathbf{w}}, \ldots, f_r^{\mathbf{w}}).$$

Now, let $F \in \text{Aut}_k k[\mathbf{x}]$. Then, the Jacobian of F is nonzero. Hence, we have $\prod_{l=1}^n (\partial f_l / \partial x_{\sigma(l)}) \neq 0$ for some $\sigma \in \mathfrak{S}_n$. Then, f_l depends on $x_{\sigma(l)}$ for each l, and so $\deg_{\mathbf{w}} f_l \geq w_{\sigma(l)}$ for each l. Thus, we obtain

$$\deg_{\mathbf{w}} F \geq w_1 + \cdots + w_n =: |\mathbf{w}|. \tag{1.1.7}$$

Proposition 1.1.5 *If the equality holds in (1.1.7), then F, $F^{\mathbf{w}} \in T(n, k)$.*

Proof   We only prove the case $n = 3$. The general case is left to the reader. Without loss of generality, we may assume that $\deg_{\mathbf{w}} f_i = w_i$ for each i, and $\mathbf{w}$ satisfies one of the following:

(1) $w_1 = w_2 = w_3$.   (2) $w_1 < w_2 < w_3$.   (3) $w_1 = w_2 < w_3$.   (4) $w_3 < w_1 = w_2$.

Then, F and $F^{\mathbf{w}}$ are tame automorphisms of the following types, where $\mathrm{Aff}(2, k)$ denotes the set of affine automorphisms of $k[x_1, x_2]$:

(1) affine automorphism.
(2) (g_1, g_2, g_3), where $g_i \in k^* x_i + k[x_1, \ldots, x_{i-1}]$ for $i = 1, 2, 3$.
(3) (G', g), where $G' \in \mathrm{Aff}(2, k)$ and $g \in k^* x_3 + k[x_1, x_2]$.
(4) $(G' + G'', g)$, where $G' \in \mathrm{Aff}(2, k)$, $G'' \in k[x_3]^2$, and $g \in k^* x_3 + k$.   □

1.1.3   Initial Algebras and Elementary Reductions

For each k-subalgebra A of $k[\boldsymbol{x}]$, we define $A^{\mathbf{w}}$ to be the k-vector space generated by $f^{\mathbf{w}}$ for $f \in A$. Then, $A^{\mathbf{w}}$ is a k-subalgebra of $k[\boldsymbol{x}]$ by (1.1.5), which we call the **initial algebra** of A.

For $h \in k[\boldsymbol{x}] \setminus \{0\}$, we have $h^{\mathbf{w}} \in A^{\mathbf{w}}$ if there exists $\phi \in A$ such that $h^{\mathbf{w}} = \phi^{\mathbf{w}}$, i.e., $\deg_{\mathbf{w}}(h - \phi) < \deg h$. The converse is also true.

Exercise 6   Show that $h^{\mathbf{w}} \in A^{\mathbf{w}}$ implies $h^{\mathbf{w}} = \phi^{\mathbf{w}}$ for some $\phi \in A$.

For each $f_1, \ldots, f_r \in k[\boldsymbol{x}]$, we have $k[f_1, \ldots, f_r]^{\mathbf{w}} \supset k[f_1^{\mathbf{w}}, \ldots, f_r^{\mathbf{w}}]$. The equality holds if $r = 1$. In general, however, it is difficult to determine the generators of the k-algebra $k[f_1, \ldots, f_r]^{\mathbf{w}}$.

Exercise 7 (Robbiano–Sweedler [103])   Let $A = k[x_1 + x_2, x_1 x_2, x_1 x_2^2]$.

(1) Show that $x_1 x_2^l \in A$ for all $l \geq 1$. $[x_1 x_2^l = x_1 x_2^{l-1}(x_1 + x_2) - x_1 x_2^{l-2} \cdot x_1 x_2]$
(2) Show that $A \cap k[x_2] = k$ and $A^{\mathbf{w}} = k + x_1 k[x_1, x_2]$ for $\mathbf{w} = (1, 0)$.

Note: The k-algebra $A^{\mathbf{w}} = k + x_1 k[x_1, x_2]$ is not finitely generated.

Now, write $\phi \in k[f_1, \ldots, f_r] \setminus \{0\}$ as $\phi = \sum_{i_1, \ldots, i_r} u_{i_1, \ldots, i_r} f_1^{i_1} \cdots f_r^{i_r}$, where $u_{i_1, \ldots, i_r} \in k$. Then, $\deg_{\mathbf{w}} \phi$ is at most

$$\delta := \max\{\deg_{\mathbf{w}} f_1^{i_1} \cdots f_r^{i_r} \mid u_{i_1, \ldots, i_r} \neq 0\}, \text{ the \textbf{apparent w-degree} of } \phi.$$

We define $\phi' := \sum' u_{i_1,\dots,i_r} (f_1^{\mathbf{w}})^{i_1} \cdots (f_r^{\mathbf{w}})^{i_r}$, where the sum $\sum'$ is taken over $i_1, \dots, i_r \geq 0$ with $\deg_{\mathbf{w}} f_1^{i_1} \cdots f_r^{i_r} = \delta$.

Remark 1.1.6

(i) $\phi' \neq 0$ if and only if $\deg_{\mathbf{w}} \phi = \delta$.
(ii) If $\phi' \neq 0$, then $\phi^{\mathbf{w}} = \phi'$, so $\phi^{\mathbf{w}}$ belongs to $k[f_1^{\mathbf{w}}, \dots, f_r^{\mathbf{w}}]$.

If $f_1^{\mathbf{w}}, \dots, f_r^{\mathbf{w}}$ are algebraically independent over k, then ϕ' is always nonzero. Hence, the following lemma holds.

Lemma 1.1.7 *Let $f_1, \dots, f_r \in k[\mathbf{x}]$ be such that $f_1^{\mathbf{w}}, \dots, f_r^{\mathbf{w}}$ are algebraically independent over k. Then, we have $k[f_1, \dots, f_r]^{\mathbf{w}} = k[f_1^{\mathbf{w}}, \dots, f_r^{\mathbf{w}}]$.*

The following notion is important in studying the elements of $\mathrm{Aut}_k\, k[\mathbf{x}]$.

Definition 1.1.8 We say that $F \in (k[\mathbf{x}] \setminus \{0\})^r$ admits an **elementary reduction** if there exists $1 \leq l \leq r$ such that $f_l^{\mathbf{w}} \in k[f_1, \dots, \widehat{f_l}, \dots, f_r]^{\mathbf{w}}$, i.e., $\deg_{\mathbf{w}}(f_l - \phi) < \deg_{\mathbf{w}} f_l$ for some $\phi \in k[f_1, \dots, \widehat{f_l}, \dots, f_r]$ (cf. Exercise 6). We call

$$F' := (f_1, \dots, f_{l-1}, f_l - \phi, f_{l+1}, \dots, f_r)$$

an **elementary reduction** of F.

Note that $\deg_{\mathbf{w}} F' < \deg_{\mathbf{w}} F$, and $F' = FE$ holds for

$$E := (x_1, \dots, x_{l-1}, x_l - \psi, x_{l+1}, \dots, x_r),$$

where $\psi \in k[x_1, \dots, \widehat{x_l}, \dots, x_r]$ is such that $\phi = F(\psi)$.

Remark 1.1.9 $F \in (k[\mathbf{x}] \setminus \{0\})^r$ admits an elementary reduction if and only if there exists an elementary automorphism E of $k[x_1, \dots, x_r]$ such that $\deg_{\mathbf{w}} FE < \deg_{\mathbf{w}} F$. Hence, by (1.1.7), $F \in \mathrm{Aut}_k\, k[\mathbf{x}]$ admits no elementary reduction if $\deg_{\mathbf{w}} F = |\mathbf{w}|$.

Proposition 1.1.10 *If $F \in \mathrm{Aut}_k\, k[\mathbf{x}]$ satisfies the following conditions, then we have $\deg_{\mathbf{w}} F > |\mathbf{w}|$, and F admits no elementary reduction:*

(e1) *$f_1^{\mathbf{w}}, \dots, f_n^{\mathbf{w}}$ are algebraically dependent over k, but any $n - 1$ of them are algebraically independent over k.*
(e2) *$f_i^{\mathbf{w}} \notin k[f_1^{\mathbf{w}}, \dots, \widehat{f_i^{\mathbf{w}}}, \dots, f_n^{\mathbf{w}}]$ for $i = 1, \dots, n$.*

Proof By (e1), we have $F^{\mathbf{w}} \notin \mathrm{Aut}_k k[\mathbf{x}]$. This implies $\deg_{\mathbf{w}} F > |\mathbf{w}|$ by Proposition 1.1.5. By Lemma 1.1.7, the last part of (e1) implies

$$k[f_1, \ldots, \widehat{f_i}, \ldots, f_n]^{\mathbf{w}} = k[f_1^{\mathbf{w}}, \ldots, \widehat{f_i^{\mathbf{w}}}, \ldots, f_n^{\mathbf{w}}]$$

for $i = 1, \ldots, n$. Hence, F admits no elementary reduction by (e2).   □

Corollary 1.1.11 *Assume that $\mathbf{w}$ is independent. If $F \in \mathrm{Aut}_k k[\mathbf{x}]$ satisfies the following conditions, then we have $\deg_{\mathbf{w}} F > |\mathbf{w}|$, and F admits no elementary reduction:*

(E1) $\deg_{\mathbf{w}} f_1, \ldots, \deg_{\mathbf{w}} f_n$ *are linearly dependent over $\mathbb{Q}$, but any $n - 1$ of them are linearly independent over $\mathbb{Q}$.*

(E2) $\deg_{\mathbf{w}} f_i \notin \sum_{j \neq i} \mathbb{Z}_{\geq 0} \deg_{\mathbf{w}} f_j$ *for $i = 1, \ldots, n$.*

Proof Since $\mathbf{w}$ is independent, (E1) is equivalent to (e1) by Exercise 5. By Exercise 3, (E2) implies (e2).   □

Finally, we discuss well-orderedness property of $\mathbf{w}$-degrees. The following exercise is essential.

Exercise 8

(1) Show that every infinite sequence of elements of $\mathbb{Z}_{\geq 0}$ has an infinite, non-decreasing subsequence.

(2) Show that every infinite sequence $(a_i)_i$ of elements of $(\mathbb{Z}_{\geq 0})^n$ has an infinite subsequence $(a_{i_l})_l$ such that $a_{i_{l+1}} - a_{i_l} \in (\mathbb{Z}_{\geq 0})^n$ for all l.

Lemma 1.1.12 $\sum_{i=1}^n \mathbb{Z}_{\geq 0} w_i$ *is a well-ordered subset of Γ.*

Proof Suppose that the lemma is false. Then, $\sum_{i=1}^n \mathbb{Z}_{\geq 0} w_i$ contains an infinite, strictly decreasing sequence $\mathbf{a} = (a_{i,1} w_1 + \cdots + a_{i,n} w_n)_{i=1}^\infty$, where $a_{i,j} \in \mathbb{Z}_{\geq 0}$. Since $w_1, \ldots, w_n > 0$, we know by Exercise 8 (2) that $\mathbf{a}$ has an infinite, non-decreasing subsequence, which is absurd.   □

Remark 1.1.13 Let A be a k-subalgebra of $k[\mathbf{x}]$, and $h \in k[\mathbf{x}] \setminus A$. Then, $\{\deg_{\mathbf{w}} f \mid f \in h + A\}$ has a least element by Lemma 1.1.12. Hence, there exists $f \in h + A$ such that $f^{\mathbf{w}} \notin A^{\mathbf{w}}$ in view of Exercise 6.

Exercise 9 In the situation of Remark 1.1.13, let $g \in h + A$ be such that $\deg_{\mathbf{w}} g > \min\{\deg_{\mathbf{w}} f \mid f \in h + A\}$. Show that $g^{\mathbf{w}}$ belongs to $A^{\mathbf{w}}$. [Take $f \in h + A$ with $\deg_{\mathbf{w}} f < \deg_{\mathbf{w}} g$. Then, $g^{\mathbf{w}} = (g - f)^{\mathbf{w}}$ and $g - f \in A$.]

By Remark 1.1.13 and Exercise 9, the following holds for each $g \in h + A$:

$$\deg_w g = \min\{\deg_w f \mid f \in h + A\} \iff g^w \notin A^w. \qquad (1.1.8)$$

1.1.4 Wildness Criterion

Assume that $n \geq 3$. Recall that $\mathcal{T}$ is the set of $F \in k[x]^3$ such that f_1, f_2, and f_3 are algebraically independent over k.

Definition 1.1.14 (Kuroda) We say that the pair $(F, G) \in \mathcal{T}^2$ satisfies the **Shestakov–Umirbaev condition** if the following conditions hold:

(SU1) $g_1 \in f_1 + kf_3^2 + kf_3$, $g_2 \in f_2 + kf_3$ and $g_3 \in f_3 + k[g_1, g_2]$.
(SU2) $\deg_w f_1 \leq \deg_w g_1$ and $\deg_w f_2 = \deg_w g_2$.
(SU3) $(g_1^w)^2 \approx (g_2^w)^s$ for some odd number $s \geq 3$.
(SU4) $\deg_w f_3 \leq \deg_w g_1$ and $f_3^w \notin k[g_1^w, g_2^w]$.
(SU5) $\deg_w g_3 < \deg_w f_3$.
(SU6) $\deg_w g_3 < \deg_w g_1 - \deg_w g_2 + \deg_w dg_1 \wedge dg_2$.

Here, $\deg_w dg_1 \wedge dg_2$ denotes the maximum among

$$\deg_w \left| \frac{\partial(g_1, g_2)}{\partial(x_i, x_j)} \right| x_i x_j \quad \text{for} \quad 1 \leq i < j \leq n. \qquad (1.1.9)$$

Exercise 10 In the situation of Definition 1.1.14, show the following:

(1) If $\deg_w f_1 = \deg_w g_1$, then $2 \deg_w f_1 = s \deg_w f_2$.
(2) If $\deg_w f_1 < \deg_w g_1$, then $s \deg_w f_2 = 4 \deg_w f_3$. [We have $g_1^w \approx (f_3^w)^2$ or $g_1^w \approx f_3^w$ by (SU1), but $g_1^w \not\approx f_3^w$ by (SU4).]

For $i = 1, 2, 3$, let $\mathcal{E}_i$ denote the set of elementary automorphisms E of $k[x_1, x_2, x_3]$ such that $E(x_j) = x_j$ for each $j \neq i$. We set $\mathcal{E} := \bigcup_{i=1}^{3} \mathcal{E}_i$.

Remark 1.1.15 If $(F, G) \in \mathcal{T}^2$ satisfies the Shestakov–Umirbaev condition, then we have the following:

(i) $G = FE_1 E_2 E_3$ for some $E_i \in \mathcal{E}_i$ by (SU1).
(ii) G is an elementary reduction of $F' := (g_1, g_2, f_3)$ by (SU1) and (SU5).
(iii) We have $\deg_w F' \geq \deg_w F$ by (SU2), but $\deg_w G < \deg_w F$ as shown later (Lemma 1.2.2 (i)).

We say that $F \in \mathscr{T}$ admits a **Shestakov–Umirbaev reduction** if (F_σ, G_σ) satisfies the Shestakov–Umirbaev condition for some $\sigma \in \mathfrak{S}_3$ and $G \in \mathscr{T}$.

The following remark follows from Exercise 10, since $\deg_{\mathbf{w}} f_1 \leq \deg_{\mathbf{w}} g_1$ by (SU2).

Remark 1.1.16 If $F \in \mathscr{T}$ admits a Shestakov–Umirbaev reduction, then $\deg_{\mathbf{w}} f_i$ and $\deg_{\mathbf{w}} f_j$ are linearly dependent over $\mathbb{Q}$ for some $i \neq j$.

Now, assume that $n = 3$. The following is the main theorem of the Shestakov–Umirbaev theory (cf. [74], Theorem 2.1).

Theorem 1.1.17 (Kuroda) *If $F \in T(3, k)$ satisfies $\deg_{\mathbf{w}} F > |\mathbf{w}|$, then F admits an elementary reduction or a Shestakov–Umirbaev reduction.*

Therefore, if $F \in \mathrm{Aut}_k k[\mathbf{x}]$ satisfies $\deg_{\mathbf{w}} F > |\mathbf{w}|$ and admits neither an elementary reduction nor a Shestakov–Umirbaev reduction, then F is wild.

We remark that Proposition 1.1.5 and Theorem 1.1.17 imply the following algorithm for deciding tameness and wildness of elements of $\mathrm{Aut}_k k[\mathbf{x}]$:

Input: $F \in \mathrm{Aut}_k k[\mathbf{x}]$.
Output: "$F \in T(3, k)$" or "$F \notin T(3, k)$."

1. If $\deg_{\mathbf{w}} F > |\mathbf{w}|$, then go to Step 2, else output "$F \in T(3, k)$."
2. If F admits an elementary or a Shestakov–Umirbaev reduction, and G is an elementary or a Shestakov–Umirbaev reduction of F, then replace F with G and return to Step 1, else output "$F \notin T(3, k)$."

Since $\deg_{\mathbf{w}} G < \deg_{\mathbf{w}} F$ in Step 2 (cf. Remark 1.1.15 (iii)), the termination of the algorithm follows from Lemma 1.1.12. We explain how to perform the following tasks (1) and (2) in Sect. 1.6 (see also [75] for a concrete algorithm):

(1) *Decide whether F admits an elementary or a Shestakov–Umirbaev reduction.*
(2) *If F admits an elementary (resp., Shestakov–Umirbaev) reduction, then construct an elementary (resp., Shestakov–Umirbaev) reduction of F.*

In the following sections, we prove Theorem 1.1.17 **when w is independent**. The following is a corollary to Theorem 1.1.17 when $\mathbf{w}$ is independent.

Corollary 1.1.18 (Kuroda) *If $F \in \mathrm{Aut}_k k[\mathbf{x}]$ satisfies (E1) and (E2) in Corollary 1.1.11 for some independent weight $\mathbf{w}$, then F is wild.*

Proof By Corollary 1.1.11, F satisfies $\deg_{\mathbf{w}} F > |\mathbf{w}|$ and admits no elementary reduction. From (E1) and Remark 1.1.16, we see that F admits no Shestakov–Umirbaev reduction. $\qquad\square$

Exercise 11 Deduce from Corollary 1.1.18 that Nagata's automorphism (1.1.3) is wild. [cf. Example 1.1.3 (2)]

1.2 Structure of the Proof

In this section, we finish the proof of Theorem 1.1.17 when $\mathbf{w}$ is independent, by assuming Lemmas 1.2.2, 1.2.11, 1.2.12, and 1.2.15, which will be proved later. In what follows, we write "deg" for "$\deg_{\mathbf{w}}$" for simplicity. We often abbreviate "Shestakov–Umirbaev" to "SU."

1.2.1 Key Propositions

Note that (SU1), (SU2), and (SU3) imply the following three conditions, respectively:

(SU1′) $g_1 \in f_1 + k[f_2, f_3]$, $g_2 \in f_2 + k[f_3]$ and $g_3 \in f_3 + k[g_1, g_2]$.
(SU2′) $\deg f_i \le \deg g_i$ for $i = 1, 2$.
(SU3′) $\deg g_2 < \deg g_1$ and $g_1^{\mathbf{w}} \notin k[g_2^{\mathbf{w}}]$.

We say that (F, G) satisfies the **weak Shestakov–Umirbaev condition** if (SU1′), (SU2′), (SU3′), (SU4), (SU5), and (SU6) hold. By definition, the SU condition implies the weak SU condition.

Remark 1.2.1

(i) (SU1′) implies that $G = F E_1' E_2' E_3'$ for some $E_i' \in \mathscr{E}_i$.
(ii) (SU3′), (SU4), (SU5), and (SU6) depend only on G and $f_3^{\mathbf{w}}$.

Exercise 12 Show the following for $(F, G) \in \mathscr{T}^2$:

(1) Assume that (F, G) satisfies (SU1′). If $E \in \mathscr{E}_1$, or if $E \in \mathscr{E}_2$ and $E(x_2) \in x_2 + k[x_3]$, then (FE, G) satisfies (SU1′).
(2) Let $F' = (f_1', f_2', f_3') \in \mathscr{T}$ be such that $\deg f_i' \le \deg g_i$ for $i = 1, 2$ and $(f_3')^{\mathbf{w}} = f_3^{\mathbf{w}}$. If (F, G) satisfies (SU2′), (SU3′), (SU4), (SU5), and (SU6), then the same holds for (F', G).

We prove the following lemma in Sect. 1.4.

Lemma 1.2.2 *If* $(F, G) \in \mathscr{T}^2$ *satisfies the weak SU condition, then the following assertions hold*:

(i) $\deg G < \deg F$.
(ii) *There exist* $E_i \in \mathscr{E}_i$ *for* $i = 1, 2$ *such that* $\deg G = \deg G E_1 = \deg G E_1 E_2$, *and* $(F, G E_1 E_2)$ *satisfies the SU condition*.

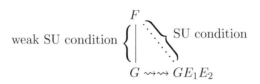

We say that $F \in \mathscr{T}$ admits a **weak Shestakov–Umirbaev reduction** if (F_σ, G_σ) satisfies the weak SU condition for some $\sigma \in \mathfrak{S}_3$ and $G \in \mathscr{T}$. We call this G a **weak Shestakov–Umirbaev reduction** of F.

Remark 1.2.3 By Lemma 1.2.2 (ii), $F \in \mathscr{T}$ admits an SU reduction if and only if F admits a weak SU reduction.

Now, assume that $n = 3$. Let $\mathscr{A}$ be the set of $F \in \mathrm{Aut}_k\, k[\boldsymbol{x}]$ for which there exist $l \geq 1$ and $F_1, \ldots, F_l \in \mathrm{Aut}_k\, k[\boldsymbol{x}]$ as follows:

① $F_1 = F$ and $\deg F_l = |\mathbf{w}|$.
② F_{i+1} is an elementary reduction or a weak SU reduction of F_i for each i.

Then, we note the following, where $F \in \mathrm{Aut}_k\, k[\boldsymbol{x}]$:

(1) If $\deg F = |\mathbf{w}|$, then F belongs to $\mathscr{A}$. Hence, $D(3, k)$ is contained in $\mathscr{A}$.
(2) $F \in \mathscr{A}$ implies $F_\sigma \in \mathscr{A}$ for any $\sigma \in \mathfrak{S}_3$. Actually, if F_{i+1} is an elementary reduction or a weak SU reduction of F_i, then $(F_{i+1})_\sigma$ is that of $(F_i)_\sigma$.
(3) If F satisfies one of the following, then F belongs to $\mathscr{A}$:

   (A1)   There exists $E \in \mathscr{E}$ such that $\deg FE < \deg F$ and $FE \in \mathscr{A}$.
   (A2) (F_σ, G) satisfies the weak SU condition for some $\sigma \in \mathfrak{S}_3$ and $G \in \mathscr{A}$.

(4) Conversely, if $F \in \mathscr{A}$ satisfies $\deg F > |\mathbf{w}|$, then (A1) or (A2) holds.
(5) (A1) implies that F admits an elementary reduction, and (A2) implies that F admits an SU reduction by Remark 1.2.3. Therefore, to prove Theorem 1.1.17, it suffices to show that $T(3, k) \subset \mathscr{A}$.

Remark 1.2.4 Remark 1.2.3 does not say that (A2) implies

(A3) (F_σ, G) satisfies the SU condition for some $\sigma \in \mathfrak{S}_3$ and $G \in \mathscr{A}$.

However, (A2) does imply (A3) under a certain hypothesis as we will see later (cf. Lemma 1.2.9 (iii)).

Exercise 13 Show that, if $F \in \mathscr{A}$ and $E \in \mathscr{E}$ satisfy $\deg FE > \deg F$, then FE belongs to $\mathscr{A}$. [FE satisfies (A1), since $E^{-1} \in \mathscr{E}$ and $(FE)E^{-1} = F \in \mathscr{A}$.]

The following is a key proposition.

Proposition 1.2.5 *If $F \in \mathscr{A}$ and $E \in \mathscr{E}$ satisfy $\deg FE \leq \deg F$, then FE belongs to $\mathscr{A}$.*

Proof of "Proposition 1.2.5 implies $T(3, k) \subset \mathscr{A}$" Proposition 1.2.5 together with Exercise 13 implies $\mathscr{A}\mathscr{E} \subset \mathscr{A}$, and so $\mathscr{A}E(3, k) \subset \mathscr{A}$. Since $D(3, k) \subset \mathscr{A}$, and $D(3, k)E(3, k) = T(3, k)$ by Exercise 2 (2), we get $T(3, k) \subset \mathscr{A}$.   □

Next, for each $F \in \mathscr{A}$, we define I_F to be the set of $i \in \{1, 2, 3\}$ for which there exists $E \in \mathscr{E}_i$ such that $\deg FE < \deg F$ and $FE \in \mathscr{A}$. By definition, F satisfies (A1) if and only if $I_F \neq \emptyset$.

The following proposition is necessary to prove Proposition 1.2.5.

Proposition 1.2.6 *Assume that $F \in \mathscr{A}$ satisfies $f_3^{\mathbf{w}} \notin k[f_2^{\mathbf{w}}]$,*

$$\deg f_1 - \deg f_2 + \deg df_1 \wedge df_2 \leq \deg f_3 < \deg f_1,$$
$$\text{and } (f_1^{\mathbf{w}})^2 \approx (f_2^{\mathbf{w}})^s \text{for some odd number } s \geq 3. \tag{1.2.1}$$

Then, we have $3 \in I_F$.

Remark 1.2.7 If $f_3^{\mathbf{w}} \in k[f_2^{\mathbf{w}}]$ or $3 \in I_F$, then we have $f_3^{\mathbf{w}} \in k[S_3]^{\mathbf{w}}$. Hence, Proposition 1.2.6 says that (1.2.1) implies $f_3^{\mathbf{w}} \in k[S_3]^{\mathbf{w}}$.

Exercise 14 Assume that $F \in \mathscr{A}$ satisfies $3 \notin I_F$, and (A3) for $\sigma = $ id. Show that

$$g_1 \notin f_1 + k[f_2] \quad \text{or} \quad g_2 \notin f_2 + k. \tag{1.2.2}$$

[If $g_1 \in f_1 + k[f_2]$ and $g_2 \in f_2 + k$, then $(f_1, f_2) = (g_1, g_2)$ by (SU1). Hence, $G \in \mathscr{A}$ is an elementary reduction of F by (SU5).]

Exercise 15 Show that $F \in \mathscr{A}$ implies $FH \in \mathscr{A}$ for each $H \in D(3, k)$. [Use induction on the number of automorphisms satisfying ① and ②.]

1.2.2 Induction

By Lemma 1.1.12, $\Delta := \{\deg F \mid F \in \mathscr{A}\}$ is a well-ordered set. We prove Propositions 1.2.5 and 1.2.6 simultaneously by induction on $\deg F$. By (1.1.7), we have $\min \Delta = |\mathbf{w}|$.

Exercise 16 Check the statements of Propositions 1.2.5 and 1.2.6 in the case where $\deg F = |\mathbf{w}|$. [Use Proposition 1.1.5 for Proposition 1.2.6.]

We fix $\mu \in \Delta$ with $\mu > |\mathbf{w}|$. To complete the induction, it suffices to prove the following claims.

Claim A If Proposition 1.2.5 holds when $\deg F < \mu$, then Proposition 1.2.6 holds when $\deg F = \mu$.

Claim B Assume that $\mathbf{w}$ is independent. If Propositions 1.2.5 and 1.2.6 hold when $\deg F < \mu$ and $\deg F \leq \mu$, respectively, then Proposition 1.2.5 holds when $\deg F = \mu$.

To prove these claims, let us first study what happens if

(†)   Proposition 1.2.5 holds when $\deg F < \mu$.

For $F, F' \in \mathscr{T}$, we write $F \rightsquigarrow F'$ if there exists $E \in \mathscr{E}$ such that $F' = FE$.

Remark 1.2.8 Assume that (†) is true. Then, for each $F \in \mathscr{A}$ with $\deg F < \mu$ and $E \in \mathscr{E}$, we have $FE \in \mathscr{A}$ whether or not $\deg FE \leq \deg F$ (cf. Exercise 13). Hence, if $F_1, \ldots, F_r \in \mathrm{Aut}_k k[\mathbf{x}]$ satisfy

$$F_1 \in \mathscr{A}, \quad F_1 \rightsquigarrow F_2 \rightsquigarrow \cdots \rightsquigarrow F_r \quad \text{and} \quad \deg F_1, \ldots, \deg F_{r-1} < \mu,$$

then $F_2, \ldots, F_r$ belong to $\mathscr{A}$.

$$\mu - - - - - - - - - - - - - \overset{\displaystyle F_6}{\nearrow} - - \quad \Rightarrow \quad F_6 \in \mathcal{A}$$
$$\mathcal{A} \ni F_1 \nearrow \quad F_2 \rightsquigarrow F_3 \searrow \quad \nearrow F_5 \nearrow$$
$$F_4$$

Using Remark 1.2.8, we can prove the following lemma.

Lemma 1.2.9 *Assume that* (†) *is true, and* $F \in \mathscr{A}$ *satisfies* $\deg F \leq \mu$.

(i) *If* $i \in I_F$ *and* $E \in \mathscr{E}_i$, *then we have* $FE \in \mathscr{A}$.
(ii) *If distinct* $i, j, l \in \{1, 2, 3\}$ *satisfy* $f_i^{\mathbf{w}} \in k[f_j^{\mathbf{w}}]$ *and* $l \in I_F$, *then* $i \in I_F$.
(iii) *If* (F, G) *satisfies the weak SU condition for some* $G \in \mathscr{A}$, *then* (F, G') *satisfies the SU condition for some* $G' \in \mathscr{A}$.

Proof

(i) Since $i \in I_F$, there exists $E_i \in \mathscr{E}_i$ such that $\deg FE_i < \deg F \leq \mu$ and $FE_i \in \mathscr{A}$. Then, we have $E_i^{-1}E \in \mathscr{E}_i$ and $FE_i \rightsquigarrow FE_i(E_i^{-1}E) = FE$.
(ii) We may assume that $(i, j, l) = (1, 2, 3)$. Then, there exist $\alpha \in k^*$, $t \geq 1$, and $\phi_3 \in k[S_3]$ such that $\deg f_1' < \deg f_1$, $\deg f_3' < \deg f_3$, and $(f_1, f_2, f_3') \in \mathscr{A}$, where $f_1' := f_1 - \alpha f_2^t$ and $f_3' := f_3 - \phi_3$. Since $(f_1, f_2, f_3') \rightsquigarrow (f_1', f_2, f_3') \rightsquigarrow (f_1', f_2, f_3)$, we have $(f_1', f_2, f_3) \in \mathscr{A}$. This implies $1 \in I_F$.
(iii) In the notation of Lemma 1.2.2 (ii), we have $G \rightsquigarrow GE_1 \rightsquigarrow GE_1E_2$, in which $\deg GE_1 = \deg G < \deg F \leq \mu$ by Lemma 1.2.2 (i).   □

The following remark is a consequence of Lemma 1.2.9 (iii).

Remark 1.2.10 Assume that (†) is true. If $F \in \mathscr{A}$ satisfies $|\mathbf{w}| < \deg F \leq \mu$, then we have $I_F \neq \emptyset$ or (A3).

In the rest of Sect. 1.2, we demonstrate Claims A and B.

1.2.3   Proof of Claim A

We need the following lemma to prove Claim A, whose proof is one of the most difficult parts of the Shestakov–Umirbaev theory (cf. Sects. 1.4 and 1.5).

Lemma 1.2.11 *Assume that* $F \in \mathscr{T}$ *satisfies* $f_3^{\mathbf{w}} \notin k[f_2^{\mathbf{w}}]$ *and* (1.2.1).

(i) *If* (F_σ, G) *satisfies the SU condition for some* $\sigma \in \mathfrak{S}_3$ *and* $G \in \mathscr{T}$, *then we have* $\sigma(3) = 3$ *and* $F_\sigma E = G$ *for some* $E \in \mathscr{E}_3$.
(ii) *If* $f_2^{\mathbf{w}}$ *belongs to* $k[S_2]^{\mathbf{w}}$, *then* $f_1^{\mathbf{w}} \approx (f_3^{\mathbf{w}})^2$.
(iii) *Assume that* $f_1^{\mathbf{w}}$ *belongs to* $k[S_1]^{\mathbf{w}}$, *and let* $f_1' \in f_1 + k[S_1]$ *be such that* $(f_1')^{\mathbf{w}} \notin k[S_1]^{\mathbf{w}}$ (*cf. Remark 1.1.13*). *Then, the following assertions hold:*
    (a) $f_2^{\mathbf{w}}$ *and* $f_3^{\mathbf{w}}$ *are algebraically dependent over* k.
    (b) $F' := (f_1', f_2, f_3)$ *admits no elementary reduction.*
    (c) *If* (F_σ', G) *satisfies the weak SU condition for some* $\sigma \in \mathfrak{S}_3$ *and* $G \in \mathscr{T}$, *then* (F, G) *satisfies the weak SU condition.*

Now, we derive Claim A from Lemma 1.2.11. Assume that (†) is true, and $F \in \mathscr{A}$ satisfies $f_3^{\mathbf{w}} \notin k[f_2^{\mathbf{w}}]$, (1.2.1), and $\deg F = \mu$. Our goal is to show that $3 \in I_F$. By Remark 1.2.10, we have (A3) or $I_F \neq \emptyset$. We demonstrate that $2 \in I_F \Rightarrow 1 \in I_F \Rightarrow$ (A3) $\Rightarrow 3 \in I_F$.

Exercise 17

(1) Derive from Lemma 1.2.11 (i) that (A3) implies $3 \in I_F$. [Since $G \in \mathscr{A}$ and $\deg G < \deg F = \deg F_\sigma$, we have $3 \in I_{F_\sigma}$.]
(2) Derive from Lemmas 1.2.9 (ii) and 1.2.11 (ii) that $2 \in I_F$ implies $1 \in I_F$. [$2 \in I_F$ implies $f_2^{\mathbf{w}} \in k[S_2]^{\mathbf{w}}$, so $f_1^{\mathbf{w}} \approx (f_3^{\mathbf{w}})^2 \in k[f_3^{\mathbf{w}}]$ by Lemma 1.2.11 (ii).]

Proof of "$1 \in I_F$ implies (A3)" If $1 \in I_F$, then there exists $f_1' \in f_1 + k[S_1]$ such that $\deg f_1' < \deg f_1$ and $F' := (f_1', f_2, f_3) \in \mathscr{A}$. We may assume that $(f_1')^{\mathbf{w}} \notin k[S_1]^{\mathbf{w}}$ by Remark 1.1.13 and Lemma 1.2.9 (i). Then, (a), (b), and (c) of Lemma 1.2.11 (iii) hold. By (a), $(F')^{\mathbf{w}}$ is not in $\mathrm{Aut}_k k[\mathbf{x}]$. This implies that $\deg_{\mathbf{w}} F' > |\mathbf{w}|$ in view of Proposition 1.1.5. Since $I_{F'} = \emptyset$ by (b), and since $F' \in \mathscr{A}$, it follows that F' satisfies (A2), i.e., (F_σ', G) satisfies the weak SU condition for some $\sigma \in \mathfrak{S}_3$ and $G \in \mathscr{A}$ (①). Then, (F, G) satisfies the weak SU condition by (c) (②). By Lemma 1.2.9 (iii), this implies that (F, G') satisfies the SU condition for some $G' \in \mathscr{A}$ (③). Therefore, F satisfies (A3).   □

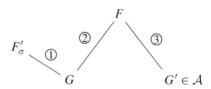

1.2.4   Proof of Claim B

To prove Claim B, take $F \in \mathscr{A}$ and $E \in \mathscr{E}$ with $\deg F = \mu$ and $\deg FE \leq \mu$. Our goal is to show that FE belongs to $\mathscr{A}$, under the assumption that $\mathbf{w}$ is independent, (†) is true, and Proposition 1.2.6 holds for automorphisms of degree at most μ. By Remark 1.2.10, we have either "(A3) and $I_F = \emptyset$" or "$I_F \neq \emptyset$."

In the former case, we may assume that F satisfies (A3) for $\sigma = \mathrm{id}$, i.e., (F, G) satisfies the SU condition for some $G \in \mathscr{A}$. Since $I_F = \emptyset$, we have (1.2.2) by Exercise 14. Then, by Lemma 1.2.12 below, (FE, G) satisfies the weak SU condition. Since G is in $\mathscr{A}$, this implies $FE \in \mathscr{A}$ by (A2).

Lemma 1.2.12 *Let $F, G \in \mathcal{F}$ and $E \in \mathcal{E}$ be such that $\deg FE \leq \deg F$, and (F, G) satisfies the weak SU condition. If $E \in \mathcal{E}_1 \cup \mathcal{E}_2$ or if $\mathbf{w}$ is independent and (1.2.2) holds, then (FE, G) satisfies the weak SU condition.*

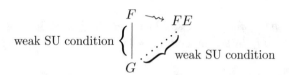

Exercise 18 Prove Lemma 1.2.12 when $E \in \mathcal{E}_1$, and when $E \in \mathcal{E}_2$ and $E(x_2) \in x_2 + k[x_3]$. [cf. Exercise 12] (The rest of the lemma is proved in Sect. 1.4.)

Next, assume that $I_F \neq \emptyset$, say $3 \in I_F$. Then, there exists $\phi_3 \in k[S_3]$ such that $G := (f_1, f_2, g_3) \in \mathcal{A}$ and $\deg g_3 < \deg f_3$, where $g_3 := f_3 - \phi_3$. We may assume that $g_3^{\mathbf{w}} \notin k[S_3]^{\mathbf{w}}$ by Remark 1.1.13 and Lemma 1.2.9 (i).

If $E \in \mathcal{E}_3$, then we have $FE \in \mathcal{A}$ by Lemma 1.2.9 (i). We assume that $E \in \mathcal{E}_1$, since the case $E \in \mathcal{E}_2$ is similar. Then, we have $FE = (f_1 - \phi_1, f_2, f_3)$ for some $\phi_1 \in k[S_1]$ with $\deg \phi_1 \leq \deg f_1$, since $\deg FE \leq \deg F$.

Remark 1.2.13 If (F, G) satisfies the weak SU condition, then FE belongs to $\mathcal{A}$ by Lemma 1.2.12 as before, whether or not (1.2.2) holds.

Exercise 19 Check that (F, G) satisfies (SU1'), (SU2'), and (SU5).

We use Remark 1.2.8 for ①, ③, and ④ of the following exercise:

$$\mu - - - \; F = (f_1, f_2, f_3) \; - - - - - - - - - -$$
$$FE = (f_1 - \phi_1, f_2, f_3)$$
$$\Rightarrow \quad FE \in \mathcal{A}$$
$$G = (f_1, f_2, f_3 - \phi_3) \in \mathcal{A}$$

Exercise 20 Show that FE belongs to $\mathcal{A}$ if one of the following holds:

① $\phi_1 \in k[f_2]$.   ② $f_1^{\mathbf{w}} \in k[f_2^{\mathbf{w}}]$.   ③ $f_3^{\mathbf{w}} \in k[f_2^{\mathbf{w}}]$.   ④ $f_1^{\mathbf{w}} \approx f_3^{\mathbf{w}}$.
[① $G \rightsquigarrow (f_1 - \phi_1, f_2, g_3) \rightsquigarrow (f_1 - \phi_1, f_2, f_3)$. ② Since $3 \in I_F$, we have $1 \in I_F$ by Lemma 1.2.9 (ii). Hence, $FE \in \mathcal{A}$ by Lemma 1.2.9 (i). ③ Take $a \in k^*$ and $l \geq 1$ with $\deg f < \deg f_3$, where $f := f_3 - af_2^l$. Then, $G' := (f_1, f_2, f)$ lies in $\mathcal{A}$ by Lemma 1.2.9 (i), and $G' \rightsquigarrow (f_1 - \phi_1, f_2, f) \rightsquigarrow (f_1 - \phi_1, f_2, f_3)$. ④ Take $a \in k^*$ with $\deg f < \deg f_1 = \deg f_3$, where $f := f_3 - af_1$. Then, $G' := (f_1, f_2, f)$ lies in $\mathcal{A}$ by Lemma 1.2.9

(i), and $G' \rightsquigarrow (f_1 + a^{-1}f, f_2, f) = (a^{-1}f_3, f_2, f) \rightsquigarrow (a^{-1}f_3, f_2, -a(f_1 - \phi_1)) =: F'$. Hence, we have $F' \in \mathscr{A}$. This implies $FE \in \mathscr{A}$ (cf. Exercise 15).]

Remark 1.2.14

(i) When $2 \in I_F$, the result of Exercise 20 with f_2 and f_3 interchanged holds.
(ii) Since 3 belongs to I_F by assumption, $f_2^{\mathbf{w}} \in k[f_1^{\mathbf{w}}]$ implies $2 \in I_F$ by Lemma 1.2.9 (ii).

Our main task is to treat the cases not covered by Exercise 20 and Remark 1.2.14. We claim that such cases are exactly the following $1^{\bigstar}$ and $2^{\bigstar}$:

$1^{\bigstar}$ $\deg f_1, \deg f_i < \deg f_j$, $f_1^{\mathbf{w}}, f_j^{\mathbf{w}} \notin k[f_i^{\mathbf{w}}]$ and $\phi_1 \notin k[f_i]$ for some $(i, j) \in \{(2, 3), (3, 2)\}$.
$2^{\bigstar}$ $\deg f_i, \deg f_3 < \deg f_j$ and $f_j^{\mathbf{w}}, f_3^{\mathbf{w}} \notin k[f_i^{\mathbf{w}}]$ for some $(i, j) \in \{(1, 2), (2, 1)\}$. (This implies $f_3^{\mathbf{w}} \notin k[f_1^{\mathbf{w}}, f_2^{\mathbf{w}}]$, since $\deg f_3 < \deg f_j$ and $f_3^{\mathbf{w}} \notin k[f_i^{\mathbf{w}}]$.)

To see this, assume that ① through ④ do not hold. If $f_2^{\mathbf{w}} \in k[f_1^{\mathbf{w}}]$, we also assume that ① through ④ with f_2 and f_3 interchanged do not hold (cf. Remark 1.2.14). Then, from ②, ③, and ④, we see that $f_i^{\mathbf{w}} \not\approx f_j^{\mathbf{w}}$ for any $i \neq j$. Since $\mathbf{w}$ is independent, this means that $\deg f_i \neq \deg f_j$ for any $i \neq j$ by (1.1.6). Let $m \in \{1, 2, 3\}$ be such that $\deg f_m = \max\{\deg f_i \mid i = 1, 2, 3\}$.

Exercise 21 Under the assumption above, show the following:

(1) If $m = 1$, then $2^{\bigstar}$ holds for $(i, j) = (2, 1)$.
(2) If $m = 3$, then $1^{\bigstar}$ holds for $(i, j) = (2, 3)$.
(3) If $m = 2$ and $f_2^{\mathbf{w}} \in k[f_1^{\mathbf{w}}]$, then $1^{\bigstar}$ holds for $(i, j) = (3, 2)$.

To complete the discussion, we would like to show the following:

(4) If $m = 2$ and $f_2^{\mathbf{w}} \notin k[f_1^{\mathbf{w}}]$, then $2^{\bigstar}$ holds for $(i, j) = (1, 2)$.

For this purpose, we need (iii) of the following lemma ((ii) and (i) are used to study the cases $1^{\bigstar}$ and $2^{\bigstar}$). This lemma is one of the consequences of the **Shestakov–Umirbaev inequality** (Theorem 1.3.3) and proved in Sect. 1.3.4.

Lemma 1.2.15 *The following assertions hold for each $F \in \mathscr{T}$:*

(i) $f_3^{\mathbf{w}} \in k[S_3]^{\mathbf{w}}$ and $2^{\bigstar}$ imply $(f_j^{\mathbf{w}})^2 \approx (f_i^{\mathbf{w}})^q$ for some odd number $q \geq 3$.
(ii) Assume that $\deg f_1, \deg f_i < \deg f_j$ and $f_j^{\mathbf{w}} \notin k[f_i^{\mathbf{w}}]$ for some $(i, j) \in \{(2, 3), (3, 2)\}$. If there exists $\phi \in k[S_1] \setminus k[f_i]$ such that $\deg \phi \leq \deg f_1$, then (f_j, f_i, f_1) satisfies (1.2.1).

(iii) *Assume that* $\deg f_3 < \deg f_2$, $f_2^{\mathbf{w}} \notin k[f_1^{\mathbf{w}}]$, *and* $f_3^{\mathbf{w}} \approx (f_1^{\mathbf{w}})^l$ *for some* $l \geq 2$. *Then,* $\deg \phi > \deg f_1$ *holds for each* $\phi \in k[S_1] \setminus k[f_2]$.

Proof of (4) It suffices to show that $f_3^{\mathbf{w}} \notin k[f_1^{\mathbf{w}}]$. If $f_3^{\mathbf{w}} \in k[f_1^{\mathbf{w}}]$, then we have $f_3^{\mathbf{w}} \approx (f_1^{\mathbf{w}})^p$ for some $p \geq 2$, since $f_3^{\mathbf{w}} \not\approx f_1^{\mathbf{w}}$ by ④. Since $\phi_1 \notin k[f_2]$ by ①, we get $\deg \phi_1 > \deg f_1$ thanks to Lemma 1.2.15 (iii), a contradiction. □

Now, let us complete the proof of Claim B. It remains only to consider the cases $1^\star$ and $2^\star$. We need Proposition 1.2.6 to treat these cases.

Case $1^\star$ Note that $1^\star$ implies the assumption of Lemma 1.2.15 (ii). Hence, $F' := (f_j, f_i, f_1)$ satisfies (1.2.1). Since $f_1^{\mathbf{w}} \notin k[f_i^{\mathbf{w}}]$ by $1^\star$, F' satisfies the assumptions of Proposition 1.2.6. Since $\deg F' = \deg F = \mu$, we get $3 \in I_{F'}$. This implies that $1 \in I_F$. Therefore, FE belongs to $\mathscr{A}$ by Lemma 1.2.9 (i).

Case $2^\star$ *with* $(i, j) = (2, 1)$ By Remark 1.2.13, it suffices to show that (F, G) satisfies the weak SU condition. By Exercise 19 and $2^\star$, the conditions other than (SU6) are satisfied. Suppose that (SU6) is false. Then, we have

$$\deg f_1 - \deg f_2 + \deg df_1 \wedge df_2 \leq \deg g_3 < \deg f_3 < \deg f_1.$$

On the other hand, since $3 \in I_F$, we have $f_3^{\mathbf{w}} \in k[S_3]^{\mathbf{w}}$. Hence, $(f_1^{\mathbf{w}})^2 \approx (f_2^{\mathbf{w}})^q$ holds for some odd number $q \geq 3$ by Lemma 1.2.15 (i). Thus, G satisfies (1.2.1). Since $\deg G < \deg F = \mu$, we may conclude that $g_3^{\mathbf{w}} \in k[S_3]^{\mathbf{w}}$ by Remark 1.2.7. This contradicts the choice of ϕ_3. Therefore, (SU6) holds true.

The case $2^\star$ with $(i, j) = (1, 2)$ is similar.

Note The reader may have noticed that the assumption "$\mathbf{w}$ is independent" appeared only in the following (a) and (b):

(a) The case $E \in \mathscr{E}_3$ of Lemma 1.2.12.
(b) The argument for deriving the conditions $1^\star$ and $2^\star$.

In fact, we will not use this assumption in the following sections, except for Remark 1.4.4 and Claim 11 in Sect. 1.4.2, which are parts of the proof of (a).

Results similar to (a) and (b) for a general weight $\mathbf{w}$ are found in [74], Proposition 4.4 and Claim 6, respectively. Here, the condition (1.2.2) in Lemma 1.2.12 is replaced with $k[f_1, f_2] \neq k[g_1, g_2]$ in Proposition 4.4 of [74]. We note that $k[f_1, f_2] \neq k[g_1, g_2]$ holds under the assumption of Exercise 14, and that $k[f_1, f_2] \neq k[g_1, g_2]$ implies (1.2.2). Hence, we may replace (1.2.2) with $k[f_1, f_2] \neq k[g_1, g_2]$ in Lemma 1.2.12.

Summary of Sect. 1.2

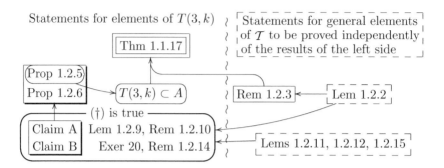

1.3   Degree Inequalities

In this section, we develop tools to analyze the elements of $\mathscr{T}$. The results in this section play crucial roles in our theory.

We use the following notation: Let y be an indeterminate over $k[\boldsymbol{x}]$. For $\Phi(y) \in k[\boldsymbol{x}][y]$ and $i \geq 0$, we denote by $\Phi^{(i)}(y)$ the ith-order derivative of $\Phi(y)$ in y. For each $f \in k[\boldsymbol{x}]$ and $i = 1, \ldots, n$, we write $f_{x_i} := \partial f / \partial x_i$.

1.3.1   Differentials

Let $\Omega_{k[\boldsymbol{x}]/k}$ be the module of differentials of $k[\boldsymbol{x}]$ over k, i.e., a free $k[\boldsymbol{x}]$-module with basis $dx_1, \ldots, dx_n$ for which we have a k-derivation

$$d : k[\boldsymbol{x}] \ni f \mapsto df := \sum_{i=1}^{n} f_{x_i} dx_i \in \Omega_{k[\boldsymbol{x}]/k}.$$

Since char $k = 0$, the following theorem holds, where $1 \leq r \leq n$.

Theorem 1.3.1 (cf. [83, §26]) $f_1, \ldots, f_r \in k[\boldsymbol{x}]$ *are algebraically independent over k if and only if $df_1, \ldots, df_r$ are linearly independent over $k[\boldsymbol{x}]$.*

Let $\bigwedge^r \Omega_{k[\boldsymbol{x}]/k}$ be the rth exterior power of the $k[\boldsymbol{x}]$-module $\Omega_{k[\boldsymbol{x}]/k}$, which is a free $k[\boldsymbol{x}]$-module with basis $dx_{i_1} \wedge \cdots \wedge dx_{i_r}$ for $1 \leq i_1 < \cdots < i_r \leq n$. For $f_1, \ldots, f_r \in k[\boldsymbol{x}]$, we have $df_1 \wedge \cdots \wedge df_r \neq 0$ if and only if $df_1, \ldots, df_r$ are linearly independent over $k[\boldsymbol{x}]$.

Hence, $df_1 \wedge \cdots \wedge df_r \neq 0$ if and only if $f_1, \ldots, f_r$ are algebraically independent over k by Theorem 1.3.1.

We can uniquely write each $\omega \in \bigwedge^r \Omega_{k[x]/k}$ as

$$\omega = \sum_{1 \leq i_1 < \cdots < i_r \leq n} f_{i_1,\ldots,i_r} dx_{i_1} \wedge \cdots \wedge dx_{i_r}, \quad \text{where} \quad f_{i_1,\ldots,i_r} \in k[x].$$

Then, we define the **w-degree** of ω by

$$\deg \omega := \max\{\deg f_{i_1,\ldots,i_r} x_{i_1} \cdots x_{i_r} \mid 1 \leq i_1 < \cdots < i_r \leq n\}.$$

By definition, we have $\deg \omega > 0$ whenever $\omega \neq 0$. Since

$$df_1 \wedge \cdots \wedge df_r = \sum_{1 \leq i_1 < \cdots < i_r \leq n} \left| \frac{\partial(f_1, \ldots, f_r)}{\partial(x_{i_1}, \ldots, x_{i_r})} \right| dx_{i_1} \wedge \cdots \wedge dx_{i_r}, \tag{1.3.1}$$

we see that "$\deg_{\mathbf{w}} dg_1 \wedge dg_2$" defined in (1.1.9) is the same as the **w**-degree of $dg_1 \wedge dg_2$ defined as above. We also note the following:

d1. $\deg df = \max\{\deg f_{x_i} x_i \mid i = 1, \ldots, n\} = \deg f$ for each $f \in k[x] \setminus k$.
d2. $\deg f\omega = \deg f + \deg \omega$ for each $f \in k[x]$ and $\omega \in \bigwedge^r \Omega_{k[x]/k}$.
d3. $\deg \omega \wedge \eta \leq \deg \omega + \deg \eta$ for each $\omega \in \bigwedge^r \Omega_{k[x]/k}$ and $\eta \in \bigwedge^s \Omega_{k[x]/k}$.

From d1 and d3, we have

$$\deg df_1 \wedge \cdots \wedge df_r \leq \deg f_1 + \cdots + \deg f_r \text{ for } f_1, \ldots, f_r \in k[x] \setminus k. \tag{1.3.2}$$

If F is an element of $\mathrm{Aut}_k k[x]$, then the Jacobian of F lies in k^*. Hence, we have $\deg df_1 \wedge \cdots \wedge df_n = |\mathbf{w}|$ by (1.3.1). Therefore, (1.1.7) is also derived from (1.3.2).

Exercise 22 Let $\phi = \Phi(f_1, \ldots, f_r) \in k[f_1, \ldots, f_r]$, where $f_1, \ldots, f_r \in k[x]$ with $r \geq 2$. Show that

$$df_1 \wedge \cdots \wedge df_{r-1} \wedge d\phi = \Phi_{x_r}(f_1, \ldots, f_r) df_1 \wedge \cdots \wedge df_r.$$

[Since d is a derivation, we have $d\phi = \sum_{i=1}^r \Phi_{x_i}(f_1, \ldots, f_r) df_i$.]

Exercise 23 Assume that $(F, G) \in \mathscr{T}^2$ satisfies the weak SU condition. Let $G' = (g_1', g_2', g_3)$, where $g_1' \in g_1 + k[g_2]$ and $g_2' \in g_2 + k$. Show the following:

(1) (F, G') satisfies (SU1'). [$g_1' \in g_1 + k[g_2] \subset g_1 + k[f_2, f_3]$ by (SU1').]
(2) If $(g_1')^{\mathbf{w}} \approx g_1^{\mathbf{w}}$, then (F, G') satisfies the weak SU condition. [We have $(g_2')^{\mathbf{w}} \approx g_2^{\mathbf{w}}$ and $dg_1' \wedge dg_2' = dg_1 \wedge dg_2$ by the choice of g_1' and g_2'.]

1.3.2   Shestakov–Umirbaev Inequality

Assume that $f_1, \ldots, f_r, g \in k[\boldsymbol{x}]$ are algebraically independent over k. Then, $\omega := df_1 \wedge \cdots \wedge df_r$ and $\omega \wedge dg$ are nonzero as remarked in Sect. 1.3.1. Take any $\Phi(y) \in k[f_1, \ldots, f_r][y] \setminus k[\boldsymbol{x}]$, and write $\Phi(y) = \sum_{i \geq 0} \phi_i y^i$, where $\phi_i \in k[f_1, \ldots, f_r]$. Then, we have

$$\deg^g \Phi := \max\{\deg \phi_i g^i \mid i \geq 0\} \geq \deg \Phi(g).$$

Since $\deg^g \Phi^{(i)} = \deg \Phi^{(i)}(g)$ for sufficiently large i, we can define

$$m^g(\Phi) := \min\{i \in \mathbb{Z}_{\geq 0} \mid \deg^g \Phi^{(i)} = \deg \Phi^{(i)}(g)\}.$$

Note that, if $\deg^g \Phi = \deg \phi_i g^i$ holds for some $i \geq 1$, then $\deg^g \Phi^{(1)} = \deg^g \Phi - \deg g$. On the other hand, we have

$$
\begin{aligned}
\deg^g \Phi > \deg \phi_i g^i \text{ for all } i \geq 1 \quad &\Rightarrow \quad \deg^g \Phi = \deg \phi_0 = \deg \Phi(g) \\
&\Rightarrow \quad m^g(\Phi) = 0.
\end{aligned}
\tag{1.3.3}
$$

Hence, we get (ii) of the following lemma. (i) is clear by definition.

Lemma 1.3.2 *If $m^g(\Phi) \geq 1$, then the following equalities hold:*
(i) $m^g(\Phi^{(1)}) = m^g(\Phi) - 1$.   (ii) $\deg^g \Phi^{(1)} = \deg^g \Phi - \deg g$.

Now, we can state and prove the **Shestakov–Umirbaev inequality** ([73], Theorem 2.1). This is a generalization of Shestakov–Umirbaev [109], Theorem 3, and plays a crucial role in our theory.

Theorem 1.3.3 (Kuroda)   *In the notation above, we have*

$$\deg \Phi(g) \geq \deg^g \Phi + m^g(\Phi)(\deg \omega \wedge dg - \deg \omega - \deg g).$$

Proof   Since $\omega \wedge d\Phi(g) = \Phi^{(1)}(g)\omega \wedge dg$ by Exercise 22, we have

$$
\begin{aligned}
\deg \omega + \deg \Phi(g) &= \deg \omega + \deg d\Phi(g) \geq \deg \omega \wedge d\Phi(g) \\
&= \deg \Phi^{(1)}(g)\omega \wedge dg = \deg \Phi^{(1)}(g) + \deg \omega \wedge dg
\end{aligned}
\tag{1.3.4}
$$

by d1, d3, and d2. We prove the theorem by induction on $m^g(\Phi)$. If $m^g(\Phi) = 0$, then $\deg \Phi(g) = \deg^g \Phi$. Hence, the assertion is true. If $m^g(\Phi) \geq 1$, then $m^g(\Phi^{(1)}) =$

$m^g(\Phi) - 1$ by Lemma 1.3.2 (i). By induction assumption, we get

$$\deg(\Phi^{(1)})(g) \geq \deg^g \Phi^{(1)} + m^g(\Phi^{(1)})(\deg \omega \wedge dg - \deg \omega - \deg g). \qquad (1.3.5)$$

The required inequality is derived from (1.3.4), (1.3.5), and Lemma 1.3.2. $\qquad \square$

Next, we define $\Phi^{w,g}(y) := \sum_{i \in I} \phi_i^w y^i$, where

$$I := \{i \mid \deg \phi_i g^i = \deg^g \Phi\}.$$

By definition, $\Phi^{w,g}(y)$ is a polynomial over $k[f_1, \ldots, f_r]^w$.

Exercise 24 Show the following:

(1) $\deg^g \Phi = \deg \Phi(g)$ if and only if $\Phi^{w,g}(g^w) \neq 0$. [cf. Remark 1.1.6 (i)]
(2) If $m^g(\Phi) \geq 1$, then $(\Phi^{w,g})^{(1)} = (\Phi^{(1)})^{w,g}$. [cf. (1.3.3)]
(3) $(\Phi^{w,g})^{(i)} = (\Phi^{(i)})^{w,g}$ holds for $i = 0, \ldots, m^g(\Phi)$. [Use induction on $m^g(\Phi)$.]

Now, by the definition of $m^g(\Phi)$, we have

$$\deg^g \Phi^{(i)} \begin{cases} > \deg \Phi^{(i)}(g) & \text{if } 0 \leq i < m^g(\Phi) \\ = \deg \Phi^{(i)}(g) & \text{if } i = m^g(\Phi). \end{cases}$$

Hence, by (3) and (1) of Exercise 24, we have

$$(\Phi^{w,g})^{(i)}(g^w) = (\Phi^{(i)})^{w,g}(g^w) \begin{cases} = 0 & \text{if } 0 \leq i < m^g(\Phi) \\ \neq 0 & \text{if } i = m^g(\Phi). \end{cases}$$

This shows that $m^g(\Phi)$ is the multiplicity of the root $y = g^w$ of $\Phi^{w,g}(y)$.

Remark 1.3.4 $m^g(\Phi)$ is determined only by $\Phi^{w,g}$ and g^w.

Let K be the field of fractions of $k[f_1, \ldots, f_r]^w$, and $p(y)$ the minimal polynomial of g^w over K. Since $\Phi^{w,g}$ lies in $K[y]$, we see that $\Phi^{w,g}$ belongs to $p(y)^{m^g(\Phi)} K[y]$. Therefore, we have $\deg_y \Phi^{w,g} \geq m^g(\Phi)[K(g^w) : K]$. This gives that

$$\deg^g \Phi \geq \deg_y \Phi^{w,g} \cdot \deg g \geq m^g(\Phi)[K(g^w) : K] \deg g. \qquad (1.3.6)$$

1.3.3   Useful Consequences

In this section, we derive useful consequences from Theorem 1.3.3.

Let $S := \{f, g\}$, where $f, g \in k[\boldsymbol{x}]$ are algebraically independent over k. Then, each $\phi \in k[S] \setminus \{0\}$ is uniquely written as $\phi = \sum_{i,j} c_{i,j} f^i g^j$, where $c_{i,j} \in k$. Let $\deg^S \phi$ denote the apparent $\mathbf{w}$-degree of ϕ (cf. Sect. 1.1.3), i.e.,

$$\deg^S \phi := \max\{\deg f^i g^j \mid i, j \in \mathbb{Z}_{\geq 0}, \ c_{i,j} \neq 0\}.$$

By Remark 1.1.6 and the discussion following it, we have $\deg \phi \leq \deg^S \phi$, and

$$f^{\mathbf{w}}, g^{\mathbf{w}} : \text{alg. indep.}/k \implies \deg \phi = \deg^S \phi \implies \phi^{\mathbf{w}} \in k[f^{\mathbf{w}}, g^{\mathbf{w}}]. \tag{1.3.7}$$

In this section, we study what happens if $\deg \phi < \deg^S \phi$.

First of all, the following lemma follows from (1.3.7) and Proposition 1.1.4.

Lemma 1.3.5 *If* $\deg \phi < \deg^S \phi$, *then* $(g^{\mathbf{w}})^p \approx (f^{\mathbf{w}})^q$ *holds for some* $p, q \geq 1$ *with* $\gcd(p, q) = 1$.

Next, we define $\Phi(y) := \sum_{i,j} c_{i,j} f^i y^j$. Then, we have $\Phi(g) = \phi$ and $\deg^g \Phi = \deg^S \phi$. Hence,

$$\deg \phi < \deg^S \phi \iff \deg \Phi(g) < \deg^g \Phi \iff m^g(\Phi) \geq 1.$$

Let $a \in k^*$ be such that $(g^{\mathbf{w}})^p = a(f^{\mathbf{w}})^q$. Then, $y^p - a(f^{\mathbf{w}})^q$ is the minimal polynomial of $g^{\mathbf{w}}$ over $k(f^{\mathbf{w}})$. Since $k(f^{\mathbf{w}})$ is equal to the field of fractions of $k[f]^{\mathbf{w}}$, we know by (1.3.6) that

$$\deg^g \Phi \geq m^g(\Phi) p \deg g. \tag{1.3.8}$$

The following lemma is a consequence of Theorem 1.3.3 and (1.3.8).

Lemma 1.3.6 *Assume that* $\deg \phi < \deg^S \phi$, *and let* $p, q \geq 1$ *be as above.*

(i) *Set* $\delta := (1/p) \deg f = (1/q) \deg g$. *Then, we have*

$$\deg \phi \geq p \deg g + \deg df \wedge dg - \deg f - \deg g$$
$$= (pq - p - q)\delta + \deg df \wedge dg. \tag{1.3.9}$$

(ii) *Let $h \in k[x]$ be such that f, g, and h are algebraically independent over k. If $h^{\mathbf{w}} = \phi^{\mathbf{w}}$, then we have*

$$\deg(h - \phi) \geq (p - 1) \deg g + \deg df \wedge dg \wedge dh - \deg df \wedge dh. \qquad (1.3.10)$$

Proof

(i) If $M := p \deg g + \deg df \wedge dg - \deg f - \deg g \leq 0$, the assertion is clear. So, assume that $M > 0$. Then, from d1, (1.3.8), and Theorem 1.3.3 with $\omega = df$, we obtain that $\deg \phi = \deg \Phi(g) \geq m^g(\Phi)M \geq M$.

(ii) Let M' be the right-hand side of (1.3.10). If $M' \leq 0$, the assertion is clear. So, assume that $M' > 0$. By assumption, we have $\deg h = \deg \phi < \deg^S \phi = \deg^g \Phi$. Hence, $\Psi(y) := h - \Phi(y)$ satisfies $\deg^g \Psi = \deg^g \Phi$ and $\Psi^{\mathbf{w},g} = \Phi^{\mathbf{w},g}$. This implies that $m^g(\Psi) = m^g(\Phi)$ (cf. Remark 1.3.4). Therefore, from d1, (1.3.8), and Theorem 1.3.3 with $\omega = df \wedge dh$, it follows that $\deg(h - \phi) = \deg \Psi(g) \geq m^g(\Phi)M' \geq M'$. □

Exercise 25 Show the following:

(1) If $f^{\mathbf{w}} \notin k[g^{\mathbf{w}}]$ and $g^{\mathbf{w}} \notin k[f^{\mathbf{w}}]$, then $p, q \geq 2$.
(2) If $p, q \geq 2$, then $pq - p - q > 0$.
(3) If $\deg f < \deg g$ and $g^{\mathbf{w}} \notin k[f^{\mathbf{w}}]$, then $2 \leq p < q$.
(4) If $2 \leq p < q$ and $pq - p - q \leq q$, then $p = 2$ and $q \geq 3$ is an odd number.
(5) If $2 \leq p < q$ and $pq - p - q \leq p$, then $p = 2$ and $q = 3$.

[Use $\gcd(p, q) = 1$ for (2), (4), and (5).]

From Exercise 25 and Lemmas 1.3.5 and 1.3.6 (i), we can derive (i), (ii), and (iii) of the following lemma.

Lemma 1.3.7 *Assume that $\deg \phi < \deg^S \phi$, and let $p, q \geq 1$ be as above.*

(i) *If $f^{\mathbf{w}} \notin k[g^{\mathbf{w}}]$ and $g^{\mathbf{w}} \notin k[f^{\mathbf{w}}]$, then we have $\deg \phi > \deg df \wedge dg$.*
(ii) *If $\deg f < \deg g$, $g^{\mathbf{w}} \notin k[f^{\mathbf{w}}]$, and $\deg \phi \leq \deg g$, then $p = 2$ and $q \geq 3$ is an odd number. Hence, we have $(g^{\mathbf{w}})^2 \approx (f^{\mathbf{w}})^q$, and*

$$\deg \phi \geq \deg g - \deg f + \deg df \wedge dg = (q - 2)\delta + \deg df \wedge dg, \text{ where } \delta := \frac{1}{2} \deg f.$$

(iii) *If $\deg f < \deg g$, $g^{\mathbf{w}} \notin k[f^{\mathbf{w}}]$, and $\deg \phi \leq \deg f$, then $(p, q) = (2, 3)$. Hence, we have $\deg \phi \geq (1/2) \deg f + \deg df \wedge dg > (1/2) \deg f$.*
(iv) *Under the assumption of (ii), we have $\deg d\phi \wedge df \geq \deg g + \deg df \wedge dg$.*

Proof of **(iv)** Since $d\phi \wedge df = \Phi^{(1)}(g)dg \wedge df$ by Exercise 22, it suffices to verify $\deg \Phi^{(1)}(g) \geq \deg g$. Since $p = 2$ by (ii), we see from Lemma 1.3.2 (ii) and (1.3.8) that $\deg^g \Phi^{(1)} \geq 2m^g(\Phi) \deg g - \deg g$. This inequality, and (1.3.5) with $\omega = df$ and $m^g(\Phi^{(1)}) = m^g(\Phi) - 1$, yields

$$\deg \Phi^{(1)}(g) \geq (m^g(\Phi) - 1)(\deg df \wedge dg - \deg f + \deg g) + \deg g.$$

Since $m^g(\Phi) \geq 1$ and $\deg f < \deg g$, we get $\deg \Phi^{(1)}(g) \geq \deg g$. $\qquad\square$

1.3.4   Exercises

The secret to mastering the Shestakov–Umirbaev theory is to learn how to use the results of Sect. 1.3.3 effectively. Here are some practical exercises.

Exercise 26   Prove Lemma 1.2.15.

Hint

(i) Take $\phi_3 \in k[S_3]$ with $\phi_3^{\mathbf{w}} = f_3^{\mathbf{w}}$. Since $2^{\bigstar}$ implies $f_3^{\mathbf{w}} \notin k[f_1^{\mathbf{w}}, f_2^{\mathbf{w}}]$, we have $\deg \phi_3 < \deg^{S_3} \phi_3$ by (1.3.7).

(ii) Since ϕ lies in $k[S_1] \setminus k[f_i]$, we have $\deg^{S_1} \phi \geq \deg f_j$. Since $\deg f_j > \deg f_1 \geq \deg \phi$, it follows that $\deg^{S_1} \phi > \deg \phi$.

(iii) Since ϕ lies in $k[S_1] \setminus k[f_2]$, we have $\deg^{S_1} \phi \geq \deg f_3 = \deg f_1^l > \deg f_1$.

Solution

(i) Since $\deg f_i < \deg f_j$, $f_j^{\mathbf{w}} \notin k[f_i^{\mathbf{w}}]$, and $\deg \phi_3 = \deg f_3 < \deg f_j$ by $2^{\bigstar}$, the assertion follows from Lemma 1.3.7 (ii).

(ii) Since $\deg f_i < \deg f_j$, $f_j^{\mathbf{w}} \notin k[f_i^{\mathbf{w}}]$, and $\deg \phi \leq \deg f_1 < \deg f_j$, there exists an odd number $s \geq 3$ such that $(f_j^{\mathbf{w}})^2 \approx (f_i^{\mathbf{w}})^s$ and

$$\deg f_1 \geq \deg \phi \geq \deg f_j - \deg f_i + \deg df_j \wedge df_i$$

by Lemma 1.3.7 (ii). By assumption, we have $\deg f_j > \deg f_1$.

(iii) Suppose that $\deg \phi \leq \deg f_1$. Then, we have $\deg^{S_1} \phi > \deg f_1 \geq \deg \phi$. Since $\deg f_3 < \deg f_2$, $f_2^{\mathbf{w}} \notin k[(f_1^{\mathbf{w}})^l] = k[f_3^{\mathbf{w}}]$, and $\deg \phi \leq \deg f_1 < \deg f_1^l = \deg f_3$, we get $\deg \phi > (1/2) \deg f_3 = (l/2) \deg f_1 \geq \deg f_1$ by Lemma 1.3.7 (iii), a contradiction. $\qquad\square$

In Exercises 27 and 28 below, $F \in \mathscr{T}$ satisfies (1.2.1). In this case, we have deg $f_1 = s\delta$, deg $f_2 = 2\delta$, and the following, where $\delta := (1/2)$ deg f_2:

$(\alpha) - \deg df_1 \wedge df_2 \geq (s - 2)\delta - \deg f_3$. (β) $(s - 2)\delta < \deg f_3 < s\delta$.

Exercise 27 Assume that $F \in \mathscr{T}$ satisfies (1.2.1), and there exists $\phi \in k[S_2]$ satisfying $\phi^{\mathbf{w}} = f_2^{\mathbf{w}}$ and deg $\phi < \deg^{S_2} \phi$. Show that $f_1^{\mathbf{w}} \approx (f_3^{\mathbf{w}})^2$.

Hint

(1) $(f_1^{\mathbf{w}})^q \approx (f_3^{\mathbf{w}})^p$ and $pq - p - q < p$ for some $p > q \geq 1$ with $\gcd(p, q) = 1$.
(2) $M := 3((p - 2)q + p)/p < 4$. [Estimate deg$(f_2 - \phi)$ using (1.3.10) and (α).]
(3) $(p, q) = (2, 1)$.

Solution

(1) Since deg $\phi < \deg^{S_2} \phi$ and deg $f_1 > \deg f_3$, the assertion follows from Lemma 1.3.5, and (1.3.9) with deg $\phi = \deg f_2 < \deg f_1$.
(2) By (1.3.10) and (α), we have

$$2\delta > \deg(f_2 - \phi) > (p - 1) \deg f_3 - \deg df_1 \wedge df_2 \geq (p - 2) \deg f_3 + (s - 2)\delta.$$

Since deg $f_3 = (q/p)$ deg $f_1 = (sq/p)\delta$, it follows that $4 > s((p-2)q+p)/p$. Since $p \geq 2$ and $s \geq 3$, this implies $M < 4$.
(3) If $q \geq 2$, then $q = 2$ and $p \geq 3$ by (1) and Exercise 25 (4). This implies $M = 3(3p - 4)/p \geq 5$. If $q = 1$ and $p \geq 3$, then $M = 3(2p - 2)/p \geq 4$.   □

Exercise 28 Assume that $F \in \mathscr{T}$ satisfies (1.2.1), and $f_i^{\mathbf{w}} \notin k[f_j^{\mathbf{w}}]$ for $(i, j) = (2, 3), (3, 2)$. Show that, if there exists $\phi \in k[S_1]$ satisfying $\phi^{\mathbf{w}} = f_1^{\mathbf{w}}$ and deg $\phi < \deg^{S_1} \phi$, then $3\delta > \deg f_1' > (7/3)\delta$ and deg $f_3 = (4/3)\delta$, where $\delta := (1/2)$ deg f_2 and $f_1' := f_1 - \phi$.

Hint

(1) $(f_2^{\mathbf{w}})^q \approx (f_3^{\mathbf{w}})^p$ for some $p, q \geq 2$ with $\gcd(p, q) = 1$.
(2) $s\delta > \deg f_1' > (N + s - 2)\delta$, where $N := 2q(p - 2)/p$. [Use (1.3.10) and (α).]
(3) $(p, q, s) = (3, 2, 3)$. [(2) implies $N < 2$.]

Solution

(1) Since deg $\phi < \deg^{S_1} \phi$ and $f_i^{\mathbf{w}} \notin k[f_j^{\mathbf{w}}]$ for $(i, j) = (2, 3), (3, 2)$, the assertion follows from Lemma 1.3.5 and Exercise 25 (1).

(2) By (1.3.10) and (α), we have

$$\deg f_1 > \deg f_1' > (p-1)\deg f_3 - \deg df_1 \wedge df_2 \geq (p-2)\deg f_3 + (s-2)\delta,$$

in which $\deg f_3 = (q/p)\deg f_2 = (2q/p)\delta$.

(3) If $p \geq 4$, then $N \geq q \geq 2$. If $p = 2$, then $\deg f_3 = q\delta$, and so $s - 2 < q < s$ by (β). This implies $q = s - 1$, and hence $\gcd(p,q) = 2$. Thus, we get $p = 3$. Then, $N = 2q/3 < 2$ yields $q = 2$. Since $(s-2)\delta < \deg f_3 = (4/3)\delta$ by (β), we get $s = 3$.   □

Exercise 29 Let $F \in \mathscr{T}$ be such that $f_1^{\mathbf{w}} \notin k[f_3^{\mathbf{w}}]$, and

$$\deg df_1 \wedge df_3, \deg f_1 > \delta, \quad \deg f_2 = 2\delta, \quad \text{and} \quad \deg f_3 = (3/2)\delta$$

for some $\delta \in \Gamma$. Show that, if there exists $\phi \in k[S_2]$ satisfying $\phi^{\mathbf{w}} = f_2^{\mathbf{w}}$ and $\deg \phi < \deg^{S_2} \phi$, then the following (1) and (2) hold:

(1) $\deg df_1 \wedge df_3 \leq (5/4)\delta$.   (2) $\deg df_2 \wedge df_3 > (1/4)\delta$.

Hint

Use (1.3.9) for (1), and (1.3.10) for (2).

Solution

By Lemma 1.3.5, $(f_1^{\mathbf{w}})^q \approx (f_3^{\mathbf{w}})^p$ holds for some $p, q \geq 1$ with $\gcd(p,q) = 1$. We claim that $p, q \geq 2$, since $f_1^{\mathbf{w}} \notin k[f_3^{\mathbf{w}}]$ and $\deg f_1^2 > 2\delta > (3/2)\delta = \deg f_3$. Since $\deg df_1 \wedge df_3 > \delta$ by assumption, (1.3.9) gives

$$2\delta = \deg f_2 = \deg \phi \geq (pq - p - q)\gamma + \deg df_1 \wedge df_3 > (pq - p - q)\gamma + \delta, \quad (1.3.11)$$

where $\gamma := (1/p)\deg f_1 = (1/q)\deg f_3$. Hence, we have $\delta > (pq - p - q)\gamma$. By assumption, $p\gamma = \deg f_1$ and $q\gamma = \deg f_3$ are greater than δ. Thus, we get $\min\{p,q\} > pq - p - q$. Therefore, $(p,q) \in \{(2,3),(3,2)\}$ by Exercise 25 (5). Since $(p/q)(3/2)\delta = (p/q)\deg f_3 = \deg f_1 > \delta$, we know that $(p,q) = (3,2)$. Then, (1) follows from (1.3.11), since $\gamma = (1/q)\deg f_3 = (3/4)\delta$. By (1.3.10), we have $2\delta > \deg(f_2 - \phi) > \deg f_1 - \deg df_2 \wedge df_3$. Since $\deg f_1 = (3/2)\deg f_3 = (9/4)\delta$, this implies (2).   □

1.3.5 Degrees of Cofactor Expansions

Let $S = \{f, g\}$ be as in Sect. 1.3.3. In Sects. 1.3.3 and 1.3.4, we discussed how to analyze $\phi \in k[S]$ with $\deg \phi < \deg^S \phi$. In this section, we develop a method to study the case where $\deg \phi = \deg^S \phi$. Precise information about $\deg^S \phi$ is useful to describe ϕ explicitly.

Remark 1.3.8 $\deg^S \phi < \deg g$ implies $\phi \in k[f]$.

Exercise 30 Assume that $\deg f = 2\delta$ and $\deg g > (s - 2)\delta$ for some $\delta \in \Gamma$ and an odd number $s \geq 3$. Show the following for $\phi \in k[S]$ with $\deg^S \phi \leq s\delta$:

(1) $\phi = ag^2 + cg + \psi$ for some $a, c \in k$ and $\psi \in k[f]$ with $\deg \psi \leq (s-1)\delta$. [$\deg fg > s\delta$ and $\deg g^l > l(s - 2)\delta \geq s\delta$ for $l \geq 3$.]
(2) If $a \neq 0$, then $s = 3$ and $\deg g \leq (3/2)\delta < \deg f$. [$s\delta \geq \deg^S \phi \geq \deg g^2 > 2(s-2)\delta$ implies that $s = 3$ and $\deg g \leq (s/2)\delta$.]

Now, let η_1, η_2, and η_3 be elements of $\Omega_{k[x]/k}$. For $i = 1, 2, 3$, we define

$$e_i := \deg \eta_i + \deg \eta_p \wedge \eta_q, \quad \text{where } 1 \leq p < q \leq 3 \text{ with } p, q \neq i.$$

The following is a key result of this section, which is a generalization of Shestakov–Umirbaev [109], Lemma 5. The proof given below is based on comparing two different cofactor expansions of a determinant (see [73], Theorem 5.2, for a more general result).

Theorem 1.3.9 *At least two of e_1, e_2, and e_3 are equal to $\max\{e_1, e_2, e_3\}$.*

Proof Write $\eta_i = \sum_{j=1}^{n} f_{i,j} dx_j$, where $f_{i,j} \in k[x]$, and set $\tilde{f}_{i,j} := f_{i,j} x_j$. Then, we have $\deg \eta_i = \max\{\deg \tilde{f}_{i,j} \mid j = 1, \ldots, n\}$, and

$$\deg \eta_p \wedge \eta_q = \max\{\deg D_{j,l}^{p,q} \mid 1 \leq j < l \leq n\}, \quad \text{where} \quad D_{j,l}^{p,q} := \begin{vmatrix} \tilde{f}_{p,j} & \tilde{f}_{p,l} \\ \tilde{f}_{q,j} & \tilde{f}_{q,l} \end{vmatrix}.$$

Now, suppose that the theorem is false. Then, by changing the indices of η_i's if necessary, we may assume that $e_1, e_2 < e_3$. We may also assume that $\deg \eta_1 \wedge \eta_2 = \deg D_{1,2}^{1,2}$ by changing the indices of x_i's if necessary. Let l be such that $\deg \eta_3 = \deg \tilde{f}_{3,l}$. Then, we have $e_3 = \deg D_{1,2}^{1,2} \tilde{f}_{3,l}$. The following are the cofactor expansions along the third column and the first row:

$$\begin{vmatrix} \tilde{f}_{1,1} & \tilde{f}_{1,2} & \tilde{f}_{1,l} \\ \tilde{f}_{2,1} & \tilde{f}_{2,2} & \tilde{f}_{2,l} \\ \tilde{f}_{3,1} & \tilde{f}_{3,2} & \tilde{f}_{3,l} \end{vmatrix} = \underbrace{D_{1,2}^{2,3} \tilde{f}_{1,l}}_{\text{w-degree } \leq e_1} - \underbrace{D_{1,2}^{1,3} \tilde{f}_{2,l}}_{\text{w-degree } \leq e_2} + \underbrace{D_{1,2}^{1,2} \tilde{f}_{3,l}}_{\text{w-degree } = e_3} \tag{1.3.12}$$

$$= \underbrace{D_{2,l}^{2,3} \tilde{f}_{1,1} - D_{1,l}^{2,3} \tilde{f}_{1,2} + D_{1,2}^{2,3} \tilde{f}_{1,l}}_{\text{w-degree } \leq e_1}. \tag{1.3.13}$$

Since $e_1, e_2 < e_3$ by supposition, the **w**-degree of (1.3.12) equals e_3, while that of (1.3.13) is at most e_1. This is a contradiction. □

Next, we derive some useful consequences from Theorem 1.3.9. For $H = (h_1, h_2, h_3) \in \mathscr{T}, a, c \in k, \Psi(h_2) \in k[h_2]$ and $\phi \in k[h_3]$, we define

$$h_1' := h_1 + ah_3^2 + ch_3 + \Psi(h_2) \quad \text{and} \quad h_2' := h_2 + \phi. \tag{1.3.14}$$

We consider the following conditions for $h_2, h_3, h_1', h_2',$ and $\Psi(h_2)$:

(a) $\deg h_1' > \deg \Psi(h_2)$ and $\deg h_2' = \deg h_2$.
(b) $\deg h_1' + \deg dh_2' \wedge dh_3 > \deg h_3 + \deg dh_1' \wedge dh_2'$.

Then, we note the following, where $\psi_1 := \Psi^{(1)}(h_2)$:

1* For $i = 1, 2$, we have $h_i' \notin k$, so $\deg dh_i' = \deg h_i'$ by d1 (cf. Sect. 1.3.1).
2* Since ϕ is in $k[h_3]$, we have $dh_2' \wedge dh_3 = dh_2 \wedge dh_3$.
3* $dh_1' \wedge dh_3 = dh_1 \wedge dh_3 + \psi_1 dh_2 \wedge dh_3$ by Exercise 22.
4* (a) implies $\deg h_1' - \deg h_2' > \deg \Psi(h_2) - \deg h_2 \geq \deg \psi_1$.
5* (b) implies $\deg h_1' + \deg dh_2' \wedge dh_3 = \deg h_2' + \deg dh_1' \wedge dh_3$ thanks to Theorem 1.3.9.

In the notation above, the following lemma holds.

Lemma 1.3.10 (a) *and* (b) *imply*

$$\deg dh_1 \wedge dh_3 = \deg dh_1' \wedge dh_3 = \deg h_1' - \deg h_2' + \deg dh_2 \wedge dh_3. \tag{1.3.15}$$

Proof By 2*, 4*, and 5*, we have

$$\deg \psi_1 dh_2 \wedge dh_3 = \deg \psi_1 dh_2' \wedge dh_3 = \deg \psi_1 + \deg dh_2' \wedge dh_3$$
$$< \deg h_1' - \deg h_2' + \deg dh_2' \wedge dh_3 = \deg dh_1' \wedge dh_3. \tag{1.3.16}$$

This implies $\deg dh_1' \wedge dh_3 = \deg dh_1 \wedge dh_3$ in view of 3*. The second equality of (1.3.15) follows from the last equality of (1.3.16) and 2*. □

Next, we consider the following conditions in addition to (a) and (b):

(c) $\phi = bh_3 + d$ for some $b, d \in k$.
(d) $\deg h_3 > \deg h_1' - \deg h_2' > 0$.

Remark 1.3.11 The conditions (a) through (d) are independent of h_1.

By (1.3.14) and (c), we have

$$dh_1' \wedge dh_2' = dh_1 \wedge dh_2 - 2ah_3dh_2 \wedge dh_3$$

$$+ b(dh_1 \wedge dh_3 + \psi_1 dh_2 \wedge dh_3) - cdh_2 \wedge dh_3. \tag{1.3.17}$$

From (1.3.15) with (d) and 4*, we see that

$$\deg h_3 + \deg dh_2 \wedge dh_3 > \deg dh_1 \wedge dh_3$$

$$> \max\{\deg dh_2 \wedge dh_3, \deg \psi_1 dh_2 \wedge dh_3\}. \tag{1.3.18}$$

We can derive the following lemma from (1.3.17) and (1.3.18).

Lemma 1.3.12 *Assume that* (a) *through* (d) *hold.*

(i) *If* $a \neq 0$ *and* $\deg dh_1' \wedge dh_2' < \deg h_3$, *then we have*

$$\deg dh_1 \wedge dh_2 = \deg h_3 + \deg dh_2 \wedge dh_3. \tag{1.3.19}$$

(ii) *If* $\deg dh_1' \wedge dh_2' < \deg dh_2 \wedge dh_3$, *then we have*

$$\deg dh_1 \wedge dh_2 = \begin{cases} \deg h_3 + \deg dh_2 \wedge dh_3 & \text{if } a \neq 0 \\ \deg dh_1 \wedge dh_3 & \text{if } a = 0 \text{ and } b \neq 0 \\ \deg dh_2 \wedge dh_3 & \text{if } a = b = 0 \text{ and } c \neq 0 \\ \deg dh_1' \wedge dh_2' & \text{if } a = b = c = 0. \end{cases}$$

(iii) *If* $\deg dh_1' \wedge dh_2' < \deg dh_2 \wedge dh_3$ *and* $\psi_1 = 0$, *then we have*

$$\deg \omega \leq \deg dh_1 \wedge dh_3, \text{ where } \omega := dh_1 \wedge dh_2 - 2ah_3dh_2 \wedge dh_3,$$

$$\deg \omega' \leq \deg dh_2 \wedge dh_3, \text{ where } \omega' := \omega + bdh_1 \wedge dh_3, \tag{1.3.20}$$

$$\deg \omega'' < \deg dh_2 \wedge dh_3, \text{ where } \omega'' := \omega' - cdh_2 \wedge dh_3.$$

Exercise 31 Let $F \in \mathscr{T}$ and $\phi \in k[S_1]$ be such that $\deg^{S_1} \phi = \deg f_1$ and

$$\deg f_1 = 3\delta, \quad \deg f_2 = 2\delta, \quad \deg f_3 = \frac{3}{2}\delta, \quad \deg df_1 \wedge df_2 \leq \frac{1}{2}\delta \tag{1.3.21}$$

for some $\delta \in \Gamma$, and let $f_1' := f_1 - \phi$. Show the following:

(1) $\phi = af_3^2 + cf_3 + ef_2 + e'$ for some $a, c, e, e' \in k$ with $a \neq 0$. [Use the result of Exercise 30 (1) with $s = 3$.]

(2) (a) through (d) and $\deg dh_1' \wedge dh_2' < \deg h_3$ hold for $H = (f_1', f_2, f_3)$ and $(h_1', h_2') = (f_1, f_2)$.

(3) $\deg df_1' \wedge df_3 = \delta + \deg df_2 \wedge df_3$ and $\deg df_1' \wedge df_2 = (3/2)\delta + \deg df_2 \wedge df_3$. [cf. Lemmas 1.3.10 and 1.3.12 (i).]

(4) If $\Phi(y) := f_1 - ay^2 - cy - ef_2 - e'$, then $\deg^{f_3} \Phi = 3\delta$ and $m^{f_3}(\Phi) \leq 1$.

(5) $\deg f_1' > \delta$. [If $\omega := df_1 \wedge df_2$, then $M := \deg \omega \wedge df_3 - \deg \omega - \deg f_3 > -2\delta$, and $\deg f_1' = \deg \Phi(f_3) \geq \deg^{f_3} \Phi + m^{f_3}(\Phi)M$ by Theorem 1.3.3.]

1.4 Weak Shestakov–Umirbaev Condition

In this section, we assume that $(F, G) \in \mathscr{F}^2$ satisfies the weak SU condition unless otherwise stated. We study detailed properties of F and G using the tools developed in Sect. 1.3. We can derive surprisingly many properties from (SU1′) through (SU6). For example, we will see that (F, G) satisfies (SU2) and (SU3) (Claims 1 and 2). We also prove Lemmas 1.2.2, 1.2.11 (i), and 1.2.12.

1.4.1 Description of g_1 and g_2

By (SU1′), we can write $g_3 = f_3 - \phi_3$, where $\phi_3 \in k[g_1, g_2]$. By (SU5), we have $\phi_3^{\mathbf{w}} = f_3^{\mathbf{w}}$. Hence, $\deg \phi_3 \leq \deg g_1$ and $\phi_3^{\mathbf{w}} \notin k[g_1^{\mathbf{w}}, g_2^{\mathbf{w}}]$ by (SU4). The latter implies $\deg \phi_3 < \deg^U \phi_3$ by (1.3.7), where $U := \{g_1, g_2\}$. Therefore, the following claim follows from (SU3′), and (ii) and (iv) of Lemma 1.3.7.

Claim 1 There exist an odd number $s \geq 3$ and $\delta \in \Gamma$ such that

$$(g_1^{\mathbf{w}})^2 \approx (g_2^{\mathbf{w}})^s, \quad \deg g_1 = s\delta, \quad \deg g_2 = 2\delta \tag{1.4.1}$$

$$\deg f_3 = \deg \phi_3 \geq (s - 2)\delta + \deg dg_1 \wedge dg_2 \tag{1.4.2}$$

$$\deg dg_2 \wedge d\phi_3 \geq \deg g_1 + \deg dg_1 \wedge dg_2. \tag{1.4.3}$$

Remark 1.4.1 If $\deg f_3 \leq \deg f_2$ ($\leq \deg g_2 = 2\delta$), then $s = 3$ by (1.4.2).

Claim 2 (i) $g_2 = f_2 + bf_3 + d$ for some $b, d \in k$. If $\deg f_2 < \deg f_3$, then $b = 0$.
(ii) $\deg f_2 = \deg g_2$ and $f_2^{\mathbf{w}} \not\approx f_3^{\mathbf{w}}$.

Proof We have $g_2 - f_2 \in k[f_3]$ by (SU1′), $\deg f_3 > \delta$ by (1.4.2) with $s \geq 3$, and $\deg(g_2 - f_2) \leq \max\{\deg f_2, \deg g_2\} = 2\delta$ by (SU2′). Hence, $g_2 - f_2 \in kf_3 + k$. Since $f_3^{\mathbf{w}} \not\approx g_2^{\mathbf{w}}$ by (SU4), the last part of (i) is clear. By (i), we have $\deg g_2 \leq \deg f_2$. Hence, we get $\deg g_2 = \deg f_2$ by (SU2′). Therefore, if $f_2^{\mathbf{w}} \approx f_3^{\mathbf{w}}$, then $g_2^{\mathbf{w}} \approx f_3^{\mathbf{w}}$ by (i). This contradicts (SU4). □

Claim 3 $f_i^{\mathbf{w}} \notin k[f_j^{\mathbf{w}}]$ for $(i, j) = (2, 3), (3, 2)$.

Proof We have $f_2^{\mathbf{w}} \not\approx f_3^{\mathbf{w}}$ by Claim 2 (ii). If $f_2^{\mathbf{w}} \approx (f_3^{\mathbf{w}})^l$ for some $l \geq 2$, then (SU2′) and (1.4.2) yield the contradiction

$$2\delta = \deg g_2 \geq \deg f_2 = \deg f_3^l > l(s - 2)\delta \geq 2\delta.$$

If $f_3^{\mathbf{w}} \approx (f_2^{\mathbf{w}})^l$ for some $l \geq 2$, then $\deg f_2 < \deg f_3$. Hence, $g_2 = f_2 + d$ by Claim 2 (i). This implies $g_2^{\mathbf{w}} = f_2^{\mathbf{w}}$. Therefore, we get $f_3^{\mathbf{w}} \approx (f_2^{\mathbf{w}})^l = (g_2^{\mathbf{w}})^l$, contradicting (SU4). □

Set $\gamma_i := \deg df_p \wedge df_q$ for $i = 1, 2, 3$, where $1 \leq p < q \leq 3$ with $p, q \neq i$.

Claim 4 $\gamma_1 = \deg dg_2 \wedge df_3 \geq \deg g_1 + \deg dg_1 \wedge dg_2$.

Proof Since $f_2 = g_2 - bf_3 - d$ and $f_3 = g_3 + \phi_3$, we have

$$df_2 \wedge df_3 = dg_2 \wedge df_3 = dg_2 \wedge dg_3 + dg_2 \wedge d\phi_3. \tag{1.4.4}$$

On the other hand, (1.4.3), (SU6), and (1.3.2) yield

$$\deg dg_2 \wedge d\phi_3 \geq \deg g_1 + \deg dg_1 \wedge dg_2 > \deg g_2 + \deg g_3 \geq \deg dg_2 \wedge dg_3.$$

Hence, the **w**-degree of (1.4.4) is equal to that of $dg_2 \wedge d\phi_3$, which is at least $\deg g_1 + \deg dg_1 \wedge dg_2$. □

Claim 5 Let ϕ be an element of $k[S_1]$. If $\deg \phi < \deg^{S_1} \phi$, then we have $\deg \phi > \deg g_1$. Hence, $\deg \phi \leq \deg g_1$ implies $\deg^{S_1} \phi = \deg \phi \leq \deg g_1$.

Proof If $\deg \phi < \deg^{S_1} \phi$, then we have $\deg \phi > \gamma_1$ by Lemma 1.3.7 (i) and Claim 3. By Claim 4, γ_1 is greater than $\deg g_1$. □

Note that $\phi_1 := g_1 - f_1$ lies in $k[S_1]$ by (SU1′), and $\deg \phi_1 \leq \deg g_1$ by (SU2′). Hence, we have $\deg^{S_1} \phi_1 \leq \deg g_1 = s\delta$ by Claim 5. Since $\deg f_2 = \deg g_2 = 2\delta$ by Claim 2 (ii), and $\deg f_3 > (s - 2)\delta$ by (1.4.2), the following claim follows from Exercise 30.

Claim 6 $g_1 = f_1 + af_3^2 + cf_3 + \psi$ for some $a, c \in k$ and $\psi \in k[f_2]$ with $\deg \psi \leq (s-1)\delta$. If $a \neq 0$, then $\deg f_3 < \deg f_2$ and $s = 3$.

Remark 1.4.2 If $\deg f_1 < \deg g_1$, then $a \neq 0$ and $g_1^{\mathbf{w}} \approx a(f_3^2)^{\mathbf{w}}$, since $g_1^{\mathbf{w}} \not\approx f_3^{\mathbf{w}}$ by (SU4), and $\deg g_1 = s\delta > \deg \psi$. Hence, we have $s = 3$ and $\deg f_3 = (1/2) \deg g_1 = (3/2)\delta$.

Exercise 32 Assume that $(F, G) \in \mathscr{T}^2$ satisfies the weak SU condition. Show that $\deg f_3 = (3/4) \deg f_2$, or $2 \deg f_1 = s \deg f_2$ for some odd number $s \geq 3$. [We have the former if $\deg f_1 < \deg g_1$ and the latter if $\deg f_1 = \deg g_1$.]

Proof of Lemma 1.2.2 (ii) Write g_1 and g_2 as in Claims 6 and 2.

(a) If $b = 0$, then we have $g_2' := f_2 \in g_2 + k$. Since ψ lies in $k[f_2] = k[g_2]$, we see that
$g_1' := f_1 + a f_3^2 + c f_3 \in g_1 + k[g_2]$.

(b) If $b \neq 0$, then $\deg f_3 \leq \deg f_2$ by Claim 2 (i). This implies $s = 3$ by Remark 1.4.1. Since $\deg \psi \leq (s - 1)\delta = 2\delta$, we can write $\psi = e f_2 + e'$, where $e, e' \in k$. Since $g_2 = f_2 + b f_3 + d$, we have $g_2' := f_2 + b f_3 \in g_2 + k$ and $g_1' := f_1 + a f_3^2 + c' f_3 \in g_1 - e g_2 + k$, where $c' := c - eb$.

Now, set $G' := (g_1', g_2, g_3)$ and $G'' := (g_1', g_2', g_3)$. Then, there exist $E_1 \in \mathscr{E}_1$ and $E_2 \in \mathscr{E}_2$ such that $G E_1 = G'$ and $G E_1 E_2 = G''$ in both cases (a) and (b). Moreover, we have $(g_1')^{\mathbf{w}} = g_1^{\mathbf{w}}$, since $\deg \psi < \deg g_1$, and $\deg f_3 \leq \deg f_2 < \deg g_1$ if $b \neq 0$. Hence, we know by Exercise 23 that (F, G'') satisfies the weak SU condition. It is easy to see that $\deg G = \deg G' = \deg G''$. □

1.4.2 Degree Estimations

The following is a key claim of this section, from which we can derive many properties of (F, G). We use the results of Sect. 1.3.5 to prove this claim.

Claim 7 The following equalities and inequalities hold:

$$\gamma_2 = (s - 2)\delta + \gamma_1 \geq 2(s - 1)\delta + \deg dg_1 \wedge dg_2. \tag{1.4.5}$$

$$\gamma_3 = \begin{cases} \deg f_3 + \gamma_1 & \text{if } a \neq 0 \\ \gamma_2 & \text{if } a = 0 \text{ and } b \neq 0 \\ \gamma_1 & \text{if } a = b = 0 \text{ and } c \neq 0 \\ \deg dg_1 \wedge dg_2 & \text{if } a = b = c = 0. \end{cases} \tag{1.4.6}$$

$$\deg f_3 + \gamma_1 > \gamma_2 > \gamma_1 \geq \deg g_1 + \deg dg_1 \wedge dg_2 > \deg dg_1 \wedge dg_2. \tag{1.4.7}$$

If ψ belongs to k, then (1.3.20) holds with $H = F$.

Proof From Claims 2 and 6, we see that (1.3.14) and (a) and (c) listed there hold for $H = F$ and $(h_1', h_2') = (g_1, g_2)$. Since $\deg g_1 \geq \deg f_3$ by (SU4), we have $\deg dg_2 \wedge df_3 \geq \deg f_3 + \deg dg_1 \wedge dg_2$ by Claim 4. This implies (b). We have (d) by (1.4.1) and (1.4.2). Thus, we can apply Lemmas 1.3.10 and 1.3.12. From (1.3.15) and Claim 4, we obtain

(1.4.5). Since $\gamma_1 > \deg dg_1 \wedge dg_2$ by Claim 4, we obtain (1.4.6) from Lemma 1.3.12 (ii). Equation (1.4.7) follows from (1.3.18) and Claim 4. The last statement follows from Lemma 1.3.12 (iii) and (1.4.7).   □

Remark 1.4.3 Equations (1.4.6) and (1.4.7) imply the following:

(i) If $\gamma_3 = \deg f_3 + \gamma_1$, then a must be nonzero. Hence, we have $s = 3$ by Claim 6, and so $2 \deg g_1 = 3 \deg g_2 = 3 \deg f_2$ by (1.4.1) and Claim 2 (ii).
(ii) If γ_3 is in $\{\gamma_2, \gamma_1, \deg dg_1 \wedge dg_2\}$, then $a = 0$. Hence, we have $\deg f_1 = \deg g_1$ by Remark 1.4.2, and so $2 \deg f_1 = s \deg f_2$ by (1.4.1) and Claim 2 (ii).
(iii) Equation (1.2.2) implies $(a, b, c) \neq (0, 0, 0)$. This implies $\gamma_3 \geq \gamma_1$.

Exercise 33 Let $F, G \in \mathscr{F}$ and $\sigma \in \mathfrak{S}_3$ be such that $\gamma_3 > \gamma_2 > \gamma_1$ and (F_σ, G) satisfies the weak SU condition. Show that one of the following holds:

(1) $\sigma = \mathrm{id}$ and $2 \deg g_1 = 3 \deg f_2$.
(2) $F_\sigma = (f_2, f_3, f_1)$ and $2 \deg f_2 = s \deg f_3$ for some odd number $s \geq 3$.

[Since $\gamma_3 > \gamma_2 > \gamma_1$, we have either $\gamma_{\sigma(3)} = \deg f_{\sigma(3)} + \gamma_{\sigma(1)} > \gamma_{\sigma(2)} > \gamma_{\sigma(1)}$ or $\gamma_{\sigma(2)} > \gamma_{\sigma(1)} > \deg dg_1 \wedge dg_2 = \gamma_{\sigma(3)}$ by (1.4.6) and (1.4.7).]

Exercise 34 Assume that $(F, G) \in \mathscr{F}^2$ satisfies the SU condition. Show that, if $\deg f_i \geq \gamma_j$ for some $i, j \in \{1, 2, 3\}$, then we have $j = 3$ and $FE = G$ for some $E \in \mathscr{E}_3$. [By (SU1), g_1 and g_2 are written as in Claims 6 and 2 with $\psi = d = 0$. Note that $\gamma_2 > \gamma_1 > \deg g_1$ by (1.4.7), and $\deg g_1 \geq \deg f_i$ for $i = 1, 2, 3$ by (SU2), (SU3), and (SU4). Hence, we have $j = 3$ and $\gamma_1 > \gamma_3$. By (1.4.6) and (1.4.7), this implies $a = b = c = 0$, that is, $(g_1, g_2) = (f_1, f_2)$.]

Proof of Lemma 1.2.11 (i) Equation (1.2.1) implies $\deg f_2 > \gamma_3$. Hence, we have $\sigma(3) = 3$ and $F_\sigma E = G$ for some $E \in \mathscr{E}_3$ by Exercise 34.   □

Since $\deg f_1 + \deg f_3 \geq \deg df_1 \wedge df_3$ by (1.3.2), it follows from (1.4.5) that

$$\deg f_1 + \deg f_3 \geq 2(s - 1)\delta + \deg dg_1 \wedge dg_2$$
$$= 2 \deg g_1 - \deg g_2 + \deg dg_1 \wedge dg_2. \tag{1.4.8}$$

Proof of Lemma 1.2.2 (i) By (SU6), the right-hand side of (1.4.8) is greater than $\deg g_1 + \deg g_3$. Hence, we have $\deg F > \deg G$ by Claim 2 (ii).   □

If $\deg f_1 < \deg g_1$, then $s = 3$ and $\deg f_3 = (3/2)\delta$ by Remark 1.4.2. Hence, (1.4.8) implies the following claim.

Claim 8 If $\deg f_1 < \deg g_1$, then $\deg f_1 \geq (5/2)\delta + \deg dg_1 \wedge dg_2$.

Now, we know that one of the following holds (cf. Claims 2 (ii) and 8, Remark 1.4.2, (SU2′), (SU3′), (SU4), and (1.4.1)):

1* $(5/2)\delta < \deg f_1 < \deg g_1 = 3\delta$, $\deg f_2 = 2\delta$, and $\deg f_3 = (3/2)\delta$.
2* $\deg f_1 = \deg g_1$, $\deg f_1 > \deg f_2$, $\deg f_1 \geq \deg f_3$, and $2 \deg f_1 = s \deg f_2$.

This implies the following claim.

Claim 9 (i) $\deg f_2 < \deg f_1$ and $\deg f_3 \leq \deg f_1$.
(ii) $f_i^{\mathbf{w}} \notin k[f_j^{\mathbf{w}}]$ for $(i, j) = (1, 2), (2, 1)$.

Exercise 35 Let $F \in \mathscr{T}$ be such that $(7/3)\delta < \deg f_1 < 3\delta$, $\deg f_2 = 2\delta$, and $\deg f_3 = (4/3)\delta$ for some $\delta \in \Gamma$. Show that (F_σ, G) does not satisfy the weak SU condition for any $\sigma \in \mathfrak{S}_3$ and $G \in \mathscr{T}$.
[Suppose the contrary. Since $\deg f_1 > \deg f_i$ for $i = 2, 3$, we know that $\sigma(1) = 1$ by Claim 9 (i). By Exercise 32, we have $\deg f_{\sigma(3)} = (3/4) \deg f_{\sigma(2)}$, or $\deg f_1 = (s/2) \deg f_{\sigma(2)}$ for some odd number $s \geq 3$, which is absurd.]

Claim 10 $f_3^{\mathbf{w}} \notin k[f_1^{\mathbf{w}}]$.

Proof In view of 1* and 2*, it suffices to show that $f_3^{\mathbf{w}} \not\approx f_1^{\mathbf{w}}$ when $\deg f_1 = \deg g_1$. If $f_3^{\mathbf{w}} \approx f_1^{\mathbf{w}}$ and $\deg f_1 = \deg g_1$, then we have $\deg g_1 = \deg f_1 = \deg f_3$. By Claim 6, this implies $g_1^{\mathbf{w}} = f_1^{\mathbf{w}} + cf_3^{\mathbf{w}} \approx f_3^{\mathbf{w}}$, contradicting (SU4). $\qquad\square$

Remark 1.4.4 If $\mathbf{w}$ is independent, then the following hold thanks to (1.1.6):

(i) Since $\deg f_3 \leq \deg f_1$ by Claim 9 (i), and $f_3^{\mathbf{w}} \not\approx f_1^{\mathbf{w}}$ by Claim 10, we have $\deg f_3 < \deg f_1$.
(ii) No $\phi \in k[f_2]$ satisfies $\deg \phi = \deg f_3$, since $f_3^{\mathbf{w}} \notin k[f_2^{\mathbf{w}}]$ by Claim 3.

The following claim is used to prove Lemma 1.2.12.

Claim 11 Assume that $\mathbf{w}$ is independent, and (1.2.2) holds. Then, we have the following for each $\phi \in k[S_3]$ with $\deg \phi \leq \deg f_3$:
(i) $\phi \in k[f_2]$. (ii) $\deg \phi < \deg f_3$. (iii) $\phi \in k$ or $g_2 \in f_2 + k$.

Proof For (i), it suffices to verify $\deg^{S_3} \phi = \deg \phi$, since $\deg \phi \leq \deg f_3 < \deg f_1$ by Remark 1.4.4 (i) (cf. Remark 1.3.8). If $\deg \phi < \deg^{S_3} \phi$, then $\deg \phi > \gamma_3$ by Claim 9 (ii) and Lemma 1.3.7 (i). Since (1.2.2) implies $\gamma_3 \geq \gamma_1$ by Remark 1.4.3 (iii), and since $\gamma_1 > \deg g_1 \geq \deg f_3$ by Claim 4 and (SU4), we get $\deg \phi > \deg f_3$, a contradiction.

Since $\deg \phi \leq \deg f_3$ by assumption, (ii) follows from (i) and Remark 1.4.4 (ii). If ϕ is not in k, then $\deg f_2 \leq \deg \phi < \deg f_3$ by (i) and (ii). By Claim 2 (i), this implies $g_2 \in f_2 + k$. $\qquad \square$

Lemma 1.4.5 *Let $F \in \mathscr{T}$ be such that $f_3^{\mathbf{w}} \notin k[f_1^{\mathbf{w}}]$, and $\deg f_1 \leq s\delta$ and $\deg f_3 > (s-2)\delta$ for some odd number $s \geq 3$ and $\delta \in \Gamma$. If $f_1^{\mathbf{w}} \in k[f_3^{\mathbf{w}}]$, then we have $s = 3$ and $f_1^{\mathbf{w}} \approx (f_3^{\mathbf{w}})^2$.*

Proof Since $f_3^{\mathbf{w}} \notin k[f_1^{\mathbf{w}}]$ and $f_1^{\mathbf{w}} \in k[f_3^{\mathbf{w}}]$, we have $f_1^{\mathbf{w}} \approx (f_3^{\mathbf{w}})^l$ for some $l \geq 2$. Then, we get $s\delta \geq \deg f_1 = \deg f_3^l > l(s-2)\delta \Rightarrow s/(s-2) > l \Rightarrow s = 3$ and $l = 2$. $\qquad \square$

Claim 12 If $f_1^{\mathbf{w}} \in k[f_3^{\mathbf{w}}]$, then $s = 3$, $f_1^{\mathbf{w}} \approx (f_3^{\mathbf{w}})^2$ and $\deg f_3 = (3/2)\delta$.

Proof Since $f_3^{\mathbf{w}} \notin k[f_1^{\mathbf{w}}]$ by Claim 10, and $\deg f_1 \leq s\delta$ and $\deg f_3 > (s-2)\delta$ by (SU2') and (1.4.2), the first two parts follow from Lemma 1.4.5. Since $\deg f_1 = \deg f_3^2$, we have $\deg f_3 = (s/2)\delta = (3/2)\delta$ in view of 1* and 2*. $\qquad \square$

Claim 13 If $\phi \in k[S_2]$ satisfies $\deg \phi \leq \deg f_2$, then ϕ belongs to $k[f_3]$.

Proof It suffices to verify $\deg^{S_2} \phi = \deg \phi$, since $\deg \phi \leq \deg f_2 < \deg f_1$ by Claim 9 (i) (cf. Remark 1.3.8). Suppose that $\deg \phi < \deg^{S_2} \phi$. If $f_1^{\mathbf{w}} \notin k[f_3^{\mathbf{w}}]$, then we have $\deg \phi > \gamma_2$ by Lemma 1.3.7 (i) and Claim 10. Since $\gamma_2 > 2(s-1)\delta \geq 4\delta$ by (1.4.5), this contradicts $\deg \phi \leq \deg f_2 = 2\delta$. If $f_1^{\mathbf{w}} \in k[f_3^{\mathbf{w}}]$, then the conclusion of Claim 12 holds. Hence, (1.3.9) and (1.4.5) yield

$$\deg f_2 \geq \deg \phi \geq \deg f_1 + \gamma_2 - \deg f_1 - \deg f_3 > 2(s-1)\delta - \frac{3}{2}\delta = \frac{5}{2}\delta,$$

a contradiction. $\qquad \square$

Proof of Lemma 1.2.12 The case $E \in \mathscr{E}_1$ was proved in Exercise 18. If $E \in \mathscr{E}_2$, then $FE = (f_1, f_2 + \phi, f_3)$ for some $\phi \in k[S_2]$. Since $\deg FE \leq \deg F$, we have $\deg \phi \leq \deg f_2$. This implies $\phi \in k[f_3]$ by Claim 13. Thus, $E(x_2) \in x_2 + k[x_3]$. Therefore, the assertion follows from Exercise 18.

Assume that $E \in \mathscr{E}_3$, $\mathbf{w}$ is independent, and (1.2.2) holds. Then, we have $FE = (f_1, f_2, f_3 + \phi)$ for some $\phi \in k[S_3]$ with $\deg \phi \leq \deg f_3$. Note that (i), (ii), and (iii) of Claim 11 hold. Set $f_3' := f_3 + \phi$. Then, we have $g_1 - f_1 \in k[f_2, f_3']$ by (i), and $g_2 - f_2 \in k[f_3']$ and $g_3 - f_3' \in k[g_1, g_2]$ by (iii). Hence, (FE, G) satisfies (SU1'). Since $(f_3')^{\mathbf{w}} = f_3^{\mathbf{w}}$ by (ii), the rest of the conditions hold by Exercise 12 (2). $\qquad \square$

1.5   Completion of the Proof

In this section, we prove (ii) and (iii) of Lemma 1.2.11. This will complete the proof of Theorem 1.1.17 when $\mathbf{w}$ is independent.

Assume that $F \in \mathscr{T}$ satisfies $f_3^{\mathbf{w}} \notin k[f_2^{\mathbf{w}}]$ and (1.2.1). Then, we have

$$\deg f_1 = s\delta, \quad \deg f_2 = 2\delta, \quad \text{and} \quad (s-2)\delta + \deg df_1 \wedge df_2 \le \deg f_3 < s\delta \qquad (1.5.1)$$

by (1.2.1), where $\delta := (1/2) \deg f_2$. First, we note the following:

$1°$ Since $\deg f_3 < \deg f_1$, we have $f_3^{\mathbf{w}} \notin k[f_1^{\mathbf{w}}]$.
$2°$ Since $f_3^{\mathbf{w}} \notin k[f_2^{\mathbf{w}}]$ and $\deg f_3^2 > 2(s-2)\delta \ge 2\delta$, we have $f_2^{\mathbf{w}} \notin k[f_3^{\mathbf{w}}]$.
$3°$ Since $\deg f_2 < \deg f_1$, we get $f_2^{\mathbf{w}} \notin k[f_1^{\mathbf{w}}, f_3^{\mathbf{w}}]$ by $2°$.
$4°$ If $f_1^{\mathbf{w}} \in k[f_3^{\mathbf{w}}]$, then we have $s = 3$ and $f_1^{\mathbf{w}} \approx (f_3^{\mathbf{w}})^2$ by $1°$ and Lemma 1.4.5.
$5°$ Since $(f_1^{\mathbf{w}})^2 \approx (f_2^{\mathbf{w}})^s$, we have $f_i^{\mathbf{w}} \notin k[f_j^{\mathbf{w}}]$ for $(i, j) \in \{(1, 2), (2, 1)\}$.
$6°$ Since $\deg f_3 > (s-2)\delta = \deg f_1 - \deg f_2$, we have $\deg f_1 < \deg f_2 f_3$.
$7°$ From $5°$ and $6°$, we see that $f_1^{\mathbf{w}} \notin k[f_2^{\mathbf{w}}, f_3^{\mathbf{w}}] \setminus k[f_3^{\mathbf{w}}]$.
$8°$ From $4°$ and $7°$, we see that $f_1^{\mathbf{w}} \in k[f_2^{\mathbf{w}}, f_3^{\mathbf{w}}]$ if and only if $f_1^{\mathbf{w}} \approx (f_3^{\mathbf{w}})^2$.

Proof of Lemma 1.2.11 (ii) By assumption, there exists $\phi \in k[S_2]$ such that $\phi^{\mathbf{w}} = f_2^{\mathbf{w}}$. Then, we have $\phi^{\mathbf{w}} \notin k[f_1^{\mathbf{w}}, f_3^{\mathbf{w}}]$ by $3°$. This implies $\deg \phi < \deg^{S_2} \phi$ by (1.3.7). Hence, we get $f_1^{\mathbf{w}} \approx (f_3^{\mathbf{w}})^2$ thanks to Exercise 27.   □

Next, we prove Lemma 1.2.11 (iii). By assumption, there exists $\phi_1 \in k[S_1]$ such that $f_1' = f_1 - \phi_1$ and $f_1^{\mathbf{w}} = \phi_1^{\mathbf{w}}$. By $8°$ and (1.3.7), we have

$$f_1^{\mathbf{w}} \not\approx (f_3^{\mathbf{w}})^2 \Rightarrow \phi_1^{\mathbf{w}} = f_1^{\mathbf{w}} \notin k[f_2^{\mathbf{w}}, f_3^{\mathbf{w}}] \Rightarrow \deg \phi_1 < \deg^{S_1} \phi_1 \Rightarrow (a). \qquad (1.5.2)$$

If $f_1^{\mathbf{w}} \approx (f_3^{\mathbf{w}})^2$, then (a) holds true, since $(f_1^{\mathbf{w}})^2 \approx (f_2^{\mathbf{w}})^s$ by (1.2.1). Hence, we have only to show (b) and (c). We prove the following claims.

Claim I   If $\deg \phi_1 < \deg^{S_1} \phi_1$, then (b) holds, and F' admits no weak SU reduction.

Claim II   If $\deg \phi_1 = \deg^{S_1} \phi_1$, then (b) and (c) hold.

To show (b), it suffices to check that $f_i^{\mathbf{w}} \notin k[f_1', f_j]^{\mathbf{w}}$ holds for $(i, j) = (2, 3), (3, 2)$, since $(f_1')^{\mathbf{w}} \notin k[S_1]^{\mathbf{w}}$ by assumption.

Proof of Claim I Since $f_2^{\mathbf{w}} \notin k[f_3^{\mathbf{w}}]$ by $2°$, and $f_3^{\mathbf{w}} \notin k[f_2^{\mathbf{w}}]$ by assumption, we have $3\delta > \deg f_1' > (7/3)\delta$, $\deg f_2 = 2\delta$, and $\deg f_3 = (4/3)\delta$ by Exercise 28. Hence, F' admits no weak SU reduction by Exercise 35.

To show (b), suppose that $f_t^{\mathbf{w}} \in k[f_1', f_s]^{\mathbf{w}}$ for some $(s, t) \in \{(2, 3), (3, 2)\}$. Note that (f_1', f_s, f_t) satisfies $2^{\bigstar}$ before Exercise 21 for $(i, j) = (2, 1)$, since $\deg f_1' > \deg f_2 > \deg f_3$, $(f_1')^{\mathbf{w}} \notin k[S_1]^{\mathbf{w}}$, and $f_t^{\mathbf{w}} \notin k[f_s^{\mathbf{w}}]$. Hence, by Lemma 1.2.15 (i), $\deg f_1' = (q/2) \deg f_s$ holds for some odd number $q \geq 3$. Since $\deg f_s$ is 2δ or $(4/3)\delta$, this contradicts $3\delta > \deg f_1' > (7/3)\delta$. $\qquad\qquad\square$

Next, assume that $\deg \phi_1 = \deg^{S_1} \phi_1$. Then, we have $f_1^{\mathbf{w}} \approx (f_3^{\mathbf{w}})^2 \in k[f_3^{\mathbf{w}}]$ by (1.5.2), and so $s = 3$ by $4°$. Hence, we have $\deg f_3 = (1/2) \deg f_1 = (3/2)\delta$. Then, from (1.5.1) with $s = 3$, we get (1.3.21). Since $f_1^{\mathbf{w}} = \phi_1^{\mathbf{w}}$, we know that $\deg^{S_1} \phi_1 = \deg f_1$. Thus, we obtain from (3) and (5) of Exercise 31 that

$1^{\star}$ $\deg d f_1' \wedge d f_3 = \delta + \deg d f_2 \wedge d f_3$,
$2^{\star}$ $\deg d f_1' \wedge d f_2 = (3/2)\delta + \deg d f_2 \wedge d f_3$,
$3^{\star}$ $\deg f_1' > \delta$.

Since $(f_1')^{\mathbf{w}} \notin k[S_1]^{\mathbf{w}}$ by assumption, we have $(f_1')^{\mathbf{w}} \not\approx f_i^{\mathbf{w}}$ for $i = 2, 3$. Since $\deg f_2 = 2\delta$ and $\deg f_3 = (3/2)\delta$, we see from $3^{\star}$ that

$4^{\star}$ $f_i^{\mathbf{w}} \notin k[(f_1')^{\mathbf{w}}, f_j^{\mathbf{w}}]$ for $(i, j) \in \{(2, 3), (3, 2)\}$.

Proof of Claim II For $i = 2, 3$, we set $S_i' := \{f_1', f_2, f_3\} \setminus \{f_i\}$. First, we show $f_2^{\mathbf{w}} \notin k[S_2']^{\mathbf{w}}$ by contradiction. Suppose that $\phi_2^{\mathbf{w}} = f_2^{\mathbf{w}}$ for some $\phi_2 \in k[S_2']$. Since $f_2^{\mathbf{w}} \notin k[(f_1')^{\mathbf{w}}, f_3^{\mathbf{w}}]$ by $4^{\star}$, we have $\deg \phi_2 < \deg^{S_2'} \phi_2$ by (1.3.7). We claim that the assumption of Exercise 29 holds for (f_1', f_2, f_3). In fact, $\deg d f_1' \wedge d f_3$, $\deg f_1' > \delta$ by $1^{\star}$ and $3^{\star}$, $\deg f_2 = 2\delta$ and $\deg f_3 = (3/2)\delta$ by (1.3.21), and $(f_1')^{\mathbf{w}} \notin k[f_3^{\mathbf{w}}]$ since $(f_1')^{\mathbf{w}} \notin k[S_1]^{\mathbf{w}}$. Therefore, we have $\deg d f_1' \wedge d f_3 \leq (5/4)\delta$ and $\deg d f_2 \wedge d f_3 > (1/4)\delta$. This contradicts $1^{\star}$.

Next, suppose that $\phi_3^{\mathbf{w}} = f_3^{\mathbf{w}}$ for some $\phi_3 \in k[S_3']$. Then, we have $\phi_3^{\mathbf{w}} \notin k[(f_1')^{\mathbf{w}}, f_2^{\mathbf{w}}]$ by $4^{\star}$. This implies that $\deg \phi_3 < \deg^{S_3'} \phi_3$ by (1.3.7). Note that $f_2^{\mathbf{w}} \notin k[(f_1')^{\mathbf{w}}]$ by $4^{\star}$, and $(f_1')^{\mathbf{w}} \notin k[f_2^{\mathbf{w}}]$ since $(f_1')^{\mathbf{w}} \notin k[S_1]^{\mathbf{w}}$. Thus, we get $\deg \phi_3 > \deg d f_1' \wedge d f_2$ by Lemma 1.3.7 (i). Since $\deg \phi_3 = \deg f_3 = (3/2)\delta$ by (1.3.21), this contradicts $2^{\star}$.

Finally, assume that (F_σ', G) satisfies the weak SU condition for some $\sigma \in \mathfrak{S}_3$ and $G \in \mathscr{T}$. Then, since $\deg d f_1' \wedge d f_2 > \deg d f_1' \wedge d f_3 > \deg d f_2 \wedge d f_3$ by $1^{\star}$ and $2^{\star}$, we see that (1) or (2) of Exercise 33 holds for F', G, and σ. Since $2 \deg f_2 = 4\delta < s'(3/2)\delta = s' \deg f_3$ for $s' \geq 3$, (2) does not hold. Hence, we have $\sigma = \mathrm{id}$ and $\deg g_1 = (3/2) \deg f_2 = 3\delta = \deg f_1$ by (1). By Exercise 12 (2), this implies that (F, G) satisfies (SU2′), (SU3′), (SU4), (SU5), and (SU6). By Exercise 12 (1), (F, G) also satisfies (SU1′). $\qquad\square$

1.6   How to Find Reductions

In this section, we explain how to perform (1) and (2) stated before Corollary 1.1.18.

1.6.1   Elementary Reduction

Let $S := \{f, g\}$, where $f, g \in k[\mathbf{x}]$ are algebraically independent over k, and let $h \in k[\mathbf{x}] \setminus k[S]$. If we can perform the following (∗), then we can perform (1) and (2) for elementary reductions:

(∗)   *Decide whether $h^{\mathbf{w}}$ belongs to $k[S]^{\mathbf{w}}$. If $h^{\mathbf{w}}$ belongs to $k[S]^{\mathbf{w}}$, then construct $\phi \in k[S]$ with $\phi^{\mathbf{w}} = h^{\mathbf{w}}$.*

For this purpose, it suffices to construct a finite number of $f_1, \ldots, f_r \in k[S]$ such that $h^{\mathbf{w}} \in k[S]^{\mathbf{w}}$ if and only if $h^{\mathbf{w}} \in \sum_{i=1}^r k f_i^{\mathbf{w}}$. Actually, if $h^{\mathbf{w}} = \sum_{i=1}^r \alpha_i f_i^{\mathbf{w}}$ holds for some $\alpha_i \in k$, then we have $\phi := \sum_{i \in J} \alpha_i f_i \in k[S]$ and $\phi^{\mathbf{w}} = h^{\mathbf{w}}$, where $J := \{i \mid \deg f_i = \deg h\}$.

(A) If $f^{\mathbf{w}}$ and $g^{\mathbf{w}}$ are algebraically independent over k, then $k[S]^{\mathbf{w}} = k[f^{\mathbf{w}}, g^{\mathbf{w}}]$ by Lemma 1.1.7. In this case, we have $h^{\mathbf{w}} \in k[S]^{\mathbf{w}}$ if and only if $h^{\mathbf{w}} \in \sum_{(i,j) \in I} k(f^i g^j)^{\mathbf{w}}$, where

$$I := \{(i, j) \in (\mathbb{Z}_{\geq 0})^2 \mid i \deg f + j \deg g = \deg h\}.$$

(B) Assume that $f^{\mathbf{w}}$ and $g^{\mathbf{w}}$ are algebraically dependent over k. Then, by Proposition 1.1.4, $(g^{\mathbf{w}})^p = \alpha(f^{\mathbf{w}})^q$ holds for some $\alpha \in k^*$ and $p, q \geq 1$ with $\gcd(p, q) = 1$. By the following remark, we may assume that $p, q \geq 2$.

Remark 1.6.1 If $p = 1$, then $\deg(g - \alpha f^q) < \deg g$. Since $k[f, g - \alpha f^q] = k[S]$, we replace g with $g - \alpha f^q$. Similarly, if $q = 1$, then we replace f with $f - \alpha^{-1} g^p$. We repeat this process as long as possible. Then, because of Lemma 1.1.12, we are finally reduced to the case (A), or (B) with $p, q \geq 2$.

Now, set $\delta := (1/p) \deg f = (1/q) \deg g$. Then, we have the following:

(a) $\Delta := \deg f + \deg g - \deg df \wedge dg < (p+q)\delta$, and $\Delta \geq 0$ by (1.3.2).
(b) $M := p \deg g - \Delta > (pq - p - q)\delta \geq \delta > 0$ by (a) and Exercise 25 (2).

Next, take $\phi \in k[S] \setminus \{0\}$, and $\Phi(y) \in k[f][y]$ with $\Phi(g) = \phi$ (cf. Sect. 1.3.3). Then, we have the following:

(c) $\deg \phi \geq \deg^S \phi - m^g(\Phi)\Delta \geq m^g(\Phi)M$ by Theorem 1.3.3 and (1.3.8), since $\deg^S \phi = \deg^g \Phi$.

(d) $\deg \phi \le \deg^S \phi$ and $\deg^S \phi \in \mathbb{Z}_{\ge 0} \deg f + \mathbb{Z}_{\ge 0} \deg g \subset \mathbb{Z}_{\ge 0} \delta$.

(e) By (b) and (d), there exists $m := \max\{m \in \mathbb{Z}_{\ge 0} \mid \deg \phi \ge mM\}$.

(f) By (b) and (c), we have $m \ge m^g(\Phi)$.

(g) By (a) and (d), $l\delta \ge \deg \phi + m\Delta$ holds for sufficiently large $l \ge 1$.

(h) Let l be as in (g). Then, by (c), we have

$$\deg^S \phi \le \deg \phi + m^g(\Phi)\Delta \le \deg \phi + m\Delta \le l\delta,$$

since $m^g(\Phi) \le m$ by (f), and $\Delta \ge 0$ by (a). Therefore, ϕ belongs to the finite dimensional k-vector space

$$V_l := \sum_{ip+jq \le l} kf^i g^j.$$

Notation For a finite dimensional k-vector subspace V of $k[x]$, we define V^w to be the k-vector subspace of $k[x]$ generated by ϕ^w for $\phi \in V$.

Remark 1.6.2 We can construct a basis $f_1, \ldots, f_r$ of V satisfying $V^w = \sum_{i=1}^r kf_i^w$ as follows: Let $t_1, \ldots, t_N$ be a finite number of monomials such that $V \subset \bigoplus_{i=1}^N kt_i$. We may assume that $\deg t_1 \le \cdots \le \deg t_N$. Then, by Gaussian elimination, we can construct a basis $f_1, \ldots, f_r$ of V for which there exists $1 \le i_1 < \cdots < i_r \le N$ such that

$$f_1 \in t_{i_1} + \sum_{j=1}^{i_1-1} kt_j, \ldots, f_r \in t_{i_r} + \sum_{j=1}^{i_r-1} kt_j.$$

Since $\deg t_1 \le \cdots \le \deg t_N$, we have $f_s^w \in t_{i_s} + \sum_{j=1}^{i_s-1} kt_j$ for each s. Hence, $f_1^w, \ldots, f_r^w$ are linearly independent over k. This implies that $f_1^w, \ldots, f_r^w$ form a basis of V^w.

Now, we are ready to explain how to perform the task $(*)$.

(i) If $\deg h$ is greater than any element of $\mathbb{Z}_{\ge 0} \delta$, then there exists no $\phi \in k[S]$ with $\deg \phi = \deg h$ because of (d). Hence, h^w does not belong to $k[S]^w$.

(ii) Assume that $\deg h$ is at most some element of $\mathbb{Z}_{\ge 0} \delta$. Then, there exists

$$m := \max\{m \in \mathbb{Z}_{\ge 0} \mid \deg h \ge mM\}$$

by (b). By (a), there also exists $l \ge 1$ such that $l\delta \ge \deg h + m\Delta$. Then, for every $\phi \in k[S]$ with $\deg \phi = \deg h$, we have $\phi \in V_l$ by (h). Since $\phi^w = h^w$ implies $\deg \phi = \deg h$, we see that $h^w \in k[S]^w$ if and only if $h^w \in V_l^w$. By Remark 1.6.2, we can construct a basis $f_1, \ldots, f_r$ of V_l such that $V_l^w = \sum_{i=1}^r kf_i^w$.

Remark 1.6.3 Set $\mu := \min\{\deg p \mid p \in h + k[S]\}$. Then, for $p \in h + k[S]$, we have $\deg p = \mu$ if and only if $p^{\mathbf{w}} \notin k[S]^{\mathbf{w}}$ by (1.1.8). Using the method for (∗), we can construct such p by the following algorithm:

1. If $h^{\mathbf{w}} \notin k[S]^{\mathbf{w}}$, then output h, else go to Step 2.
2. Find $\phi \in k[S]$ with $\phi^{\mathbf{w}} = h^{\mathbf{w}}$, replace h with $h - \phi$, and return to Step 1.

Since $\deg(h - \phi) < \deg h$ in Step 2, the termination of the algorithm is guaranteed by Lemma 1.1.12.

1.6.2   Shestakov–Umirbaev Reduction

In this section, we explain how to perform (1) and (2) for SU reductions.

Remark 1.6.4

(i) If $(F, G) \in \mathscr{T}^2$ satisfies the SU condition, then the following statements hold:

  - For any $g_3' \in f_3 + k[g_1, g_2]$ with $\deg g_3' \le \deg g_3$, the pair (F, G') satisfies the SU condition, where $G' := (g_1, g_2, g_3')$.
  - We have $g_1 = f_1 + af_3^2 + cf_3$ and $g_2 = f_2 + bf_3$ for some $a, b, c \in k$ by (SU1), for which (1.3.20) holds with $H = F$ by the last part of Claim 7.
  - $\deg f_3 + \gamma_1 > \gamma_2 > \gamma_1$ by (1.4.7).

(ii) Let $F \in \mathscr{T}$ be given, and assume that $\deg f_3 + \gamma_1 > \gamma_2 > \gamma_1$. Then, $a, b, c \in k$ satisfying (1.3.20) with $H = F$ are uniquely determined from F if exist. In fact, for $\eta_1, \eta_2 \in \bigwedge^2 \Omega_{k[x]/k}$ with $\eta_2 \ne 0$, there exists at most one $\alpha \in k$ satisfying $\deg(\eta_1 + \alpha\eta_2) < \deg \eta_2$.

By Remark 1.6.4, we have the following algorithm:

Input: $F \in \mathscr{T}$.
Output: An element G of $\mathscr{T}$ such that (F, G) satisfies the SU condition, or "No such G exists" (N, for short).

1. If $\deg f_3 + \gamma_1 > \gamma_2 > \gamma_1$, then go to Step 2, else output N.
2. If there exist $a, b, c \in k$ which satisfy (1.3.20) for $H = F$, then set $g_1 := f_1 + af_3^2 + cf_3$ and $g_2 := f_2 + bf_3$, and go to Step 3, else output N.
3. Take any $g_3 \in f_3 + k[g_1, g_2]$ with $\deg g_3 = \min\{\deg h \mid h \in f_3 + k[g_1, g_2]\}$ (cf. Remark 1.6.3), and set $G := (g_1, g_2, g_3)$. If (F, G) satisfies the SU condition, then output G, else output N.

Applying this algorithm to F_σ for all $\sigma \in \mathfrak{S}_3$, we can perform (1) and (2) for SU reductions.

Counterexamples to Hilbert's Fourteenth Problem

2

2.1 Introduction

Let k be a field, $k[x] = k[x_1, \ldots, x_n]$ the polynomial ring in n variables over k, and $k(x)$ the field of fractions of $k[x]$. Take a subfield L of $k(x)$ containing k, and set $A := L \cap k[x]$. Then, *Hilbert's fourteenth problem* is stated as follows (see [93]).

Hilbert's fourteenth problem Is the k-algebra A finitely generated?

Due to Zariski [134], the answer to Hilbert's fourteenth problem is affirmative if $\mathrm{tr.deg}_k L \leq 2$. Hence, there exists no counterexample if $n \leq 2$. In 1958, Nagata [91, 92] gave the first counterexample in the case $\mathrm{tr.deg}_k L = 4$ and $n = 32$. He then posed two problems. One of them asks whether there exists a counterexample to Hilbert's fourteenth problem in the case $\mathrm{tr.deg}_k L = 3$ ([91], Problem 2).

In 1990, Roberts [104] gave a new counterexample in the case $\mathrm{tr.deg}_k L = 6$ and $n = 7$. Later, the results of Nagata and Roberts were generalized by several authors (cf., e.g. [90], [112], and [115] for generalizations of Nagata's result, and [18], [47], [64], and [68] for generalizations of Roberts' result), and various counterexamples have been constructed so far.

Finite generation of invariant rings is of great interest in the study of Hilbert's fourteenth problem. In recent research, however, finite generation of the kernels of derivations has also received attention. Let B be a k-domain, and D a k-**derivation** of B, i.e., a k-linear map $D : B \to B$ which satisfies $D(ab) = D(a)b + aD(b)$ for each $a, b \in B$. Then, the kernel

$$B^D := \{b \in B \mid D(b) = 0\}$$

A. van den Essen et al., *Polynomial Automorphisms and the Jacobian Conjecture*, Frontiers in Mathematics, https://doi.org/10.1007/978-3-030-60535-3_2

of D is a k-subalgebra of B. If $Q(B^D)$ is the field of fractions of B^D, then we can write $B^D = Q(B^D) \cap B$. Hence, the following question is a special case of Hilbert's fourteenth problem (see [117], Question 1.2.21, or [98], §4.2).

Question 2.1.1 Let D be a k-derivation of $k[\boldsymbol{x}]$. Is the k-algebra $k[\boldsymbol{x}]^D$ finitely generated?

It is well known that the answer to Question 2.1.1 is affirmative if $\operatorname{char} k > 0$ (Exercise 41). If $\operatorname{char} k = 0$, then we have $\operatorname{tr.deg}_k k[\boldsymbol{x}]^D < n$ whenever $D \neq 0$. Therefore, if $n \leq 3$, then the k-algebra $k[\boldsymbol{x}]^D$ is always finitely generated thanks to the result of Zariski [134] mentioned above. Question 2.1.1 was answered in the negative by Derksen [27], who described Nagata's counterexample as the kernel of a k-derivation.

When [117] was published in 2000, the answers to Hilbert's fourteenth problem and Question 2.1.1 were known to be negative if $n \geq 5$ by Daigle-Freudenburg [18] (see also [117], §9.6). They gave a concrete k-derivation of $k[\boldsymbol{x}]$ for $n = 5$ whose kernel is not finitely generated. We should mention that Freudenburg [47] first constructed a similar example for $n = 6$ by making use of Roberts' counterexample. Then, Daigle-Freudenburg derived the example from Freudenburg's example.

A few years later, the author of this chapter constructed counterexamples to Hilbert's fourteenth problem for $n = 4$ in [69], and finally for $n = 3$ in [70], by refining the method of [68]. This method also has its roots in Roberts' work. The counterexample given in [69] is the first counterexample in the case $\operatorname{tr.deg}_k L = 3$. The author thus solved Nagata's problem mentioned at the beginning of this chapter. He also showed in [71] that the counterexamples in [69] can be described as the kernels of k-derivations, and answered Question 2.1.1 in the negative when $n = 4$.

In this chapter, we give a powerful method for constructing counterexamples to Hilbert's fourteenth problem. Using this method, one can easily produce various kinds of counterexamples, including those in the lowest dimensions mentioned above.

The main result of this chapter is based on [76].

Note A k-derivation D of B is said to be **locally nilpotent** if, for each $b \in B$, there exists $N > 0$ such that $D^N(b) = 0$. Deveney-Finston [32] remarked that Roberts' counterexample is realized as the kernel of a locally nilpotent k-derivation of $k[\boldsymbol{x}]$. The k-derivations of Freudenburg [47] and Daigle-Freudenburg [18] are also locally nilpotent. When $n = 4$, it remains an open question whether there exists a locally nilpotent k-derivation of $k[\boldsymbol{x}]$ whose kernel is not finitely generated.

Notation In what follows, let k be any field of characteristic zero unless stated otherwise. For each integral domain R, we denote by $Q(R)$ the field of fractions of R. Let $k[\boldsymbol{x}, z] = k[x_1, \ldots, x_n, z]$ be the polynomial ring in $n + 1$ variables over k, and $k(\boldsymbol{x}, z) := Q(k[\boldsymbol{x}, z])$. For each k-domain B, we denote by $\operatorname{Aut}_k B$ the automorphism group of B as a k-algebra, and regard $\operatorname{Aut}_k B$ as a subgroup of $\operatorname{Aut}_k Q(B)$ in a natural way. For each subgroup H of $\mathrm{Aut}_k\, B$, we define the invariant ring B^H by

$$B^H := \{b \in B \mid \sigma(b) = b \text{ for all } \sigma \in H\}.$$

Note that $B^H = Q(B)^H \cap B$ if we regard H as a subgroup of $\mathrm{Aut}_k\, Q(B)$. We write $B_f := B[1/f]$ for $f \in B \setminus \{0\}$, e.g.,

$$k[\boldsymbol{x}]_{x_1} = k[x_1, \ldots, x_n, x_1^{-1}] \quad \text{and} \quad k[\boldsymbol{x}, z]_{x_1} = k[x_1, \ldots, x_n, z, x_1^{-1}].$$

If $\phi : X \to Y$ is a map, we often write $Z^\phi := \phi(Z)$ for $Z \subset X$.

2.2   Main Theorem

We have a "factory" that produces counterexamples to Hilbert's fourteenth problem. The "material" is a field $k \subset M \subset k(\boldsymbol{x})$ with certain conditions. For a given M, we construct $\theta \in \mathrm{Aut}_k\, k[\boldsymbol{x}, z]_{x_1}$ by following a simple procedure. Then, the obtained k-algebra $M(z)^\theta \cap k[\boldsymbol{x}, z]$ is not finitely generated. In other words, $M(z)^\theta$ is a counterexample to Hilbert's fourteenth problem, which is our "product."

$$M \quad \Rightarrow \quad \boxed{\text{construction of } \theta} \quad \Rightarrow \quad M(z)^\theta.$$

2.2.1   Construction

Let M be a subfield of $k(\boldsymbol{x})$ containing k which satisfies the following two conditions:

(A) $R := M \cap k[\boldsymbol{x}]$ is equal to $M \cap k[\boldsymbol{x}]_{x_1}$.
(B) There does not exist $p \in k[x_1]$ such that $R^\epsilon = k[p]$, where $\epsilon : k[\boldsymbol{x}] \to k[x_1]$ is the substitution map defined by $x_2, \ldots, x_n \mapsto 0$.

Remark 2.2.1 (B) holds if and only if R^ϵ is not integrally closed (Exercise 38 (3)).

Example 2.2.2 Assume that $n = 2$, and let $y := x_2 - x_1 + x_1^2$. Then, we have $k[\boldsymbol{x}] = k[x_1, y]$. Hence, x_1 and y form a system of variables of $k[\boldsymbol{x}]$. Set

$$s := x_1 + y, \quad t := x_1 y \quad \text{and} \quad M := k(s, t).$$

Since M is the field of symmetric rational functions in x_1 and y, we have $R := M \cap k[\boldsymbol{x}] = k[s, t]$ and

$$M \cap k[\boldsymbol{x}]_{x_1} = M \cap k[\boldsymbol{x}]_y = M \cap k[\boldsymbol{x}]_{x_1} \cap k[\boldsymbol{x}]_y = M \cap k[\boldsymbol{x}]. \qquad (2.2.1)$$

Hence, M satisfies (A). Next, observe that

$$R^\epsilon = k[\epsilon(s), \epsilon(t)] = k[x_1^2, -x_1^2 + x_1^3] = k[x_1^2, x_1^3].$$

It is easy to check that $k[x_1^2, x_1^3] \neq k[p]$ for any $p \in k[x_1]$. Therefore, (B) is satisfied.

Remark 2.2.3

(i) Let L be a subfield of $k(x)$ containing k. If tr.deg$_k$ $L \leq 1$, then there exists $p \in k[x]$ such that $L \cap k[x] = k[p]$ (Exercise 37, or [117], §1.2, Exercise 13, when $n = 1$).
(ii) By (i), $Q(R^\epsilon) \cap k[x_1] = k[p]$ holds for some $p \in k[x_1]$.
(iii) By (ii) and (B), we have $R^\epsilon \subsetneqq Q(R^\epsilon) \cap k[x_1]$.

By Remark 2.2.3 (iii), $Q(R^\epsilon) \cap k[x_1] \setminus R^\epsilon$ is not empty. Take any

$$h = h(x_1) \in Q(R^\epsilon) \cap k[x_1] \setminus R^\epsilon \tag{2.2.2}$$

and $\mathbf{t} = (t_2, \ldots, t_n) \in \mathbb{Z}^{n-1}$. Then, we define $\theta_{\mathbf{t}}^h \in \text{Aut}_k\, k[x, z]_{x_1}$ by

$$\theta_{\mathbf{t}}^h(x_1) = x_1^{-1}$$

$$\theta_{\mathbf{t}}^h(x_i) = x_1^{t_i} x_i \quad \text{for} \quad i = 2, \ldots, n \tag{2.2.3}$$

$$\theta_{\mathbf{t}}^h(z) = z + h(x_1^{-1}) = z + \theta_{\mathbf{t}}^h(h).$$

We write $\theta := \theta_{\mathbf{t}}^h$ for simplicity.

In the notation above, the main result of this chapter (Theorem 2.2.4) says that, for *sufficiently large* $t_2, \ldots, t_n$, the image $M(z)^\theta$ is a counterexample to Hilbert's fourteenth problem.

2.2.2 A Sufficient Condition on $t_2, \ldots, t_n$

We give a sufficient condition on $t_2, \ldots, t_n$ for the k-algebra $M(z)^\theta \cap k[x, z]$ to be non-finitely generated. For this purpose, we first note the following properties of θ:

(1°) If a monomial $x_1^{i_1} \cdots x_n^{i_n} \in \bigcup_{i=2}^n x_i k[x]$ satisfies $t_2, \ldots, t_n > i_1$, then

$$\theta(x_1^{i_1} \cdots x_n^{i_n}) = x_1^{-i_1}(x_1^{t_2} x_2)^{i_2} \cdots (x_1^{t_n} x_n)^{i_n} \text{ belongs to } x_1 k[x],$$

since at least one of $i_2, \ldots, i_n$ is positive.

($2°$) Assume that $p \in \ker \epsilon = \sum_{i=2}^{n} x_i k[\boldsymbol{x}]$ satisfies $t_2, \ldots, t_n > \deg_{x_1} p$. Then, by ($1°$), we have $\theta(p) \in x_1 k[\boldsymbol{x}]$.

Now, since h lies in $Q(R^\epsilon)$, there exist $f, g \in R$ such that $\epsilon(f)/\epsilon(g) = h$. By the assumption that $h \notin R^\epsilon$, we have $\epsilon(g) \notin k$. Hence, $k[x_1]$ is integral over $k[\epsilon(g)]$ (Exercise 38 (1)). Since $\epsilon(f)$ is in $k[x_1]$, we know that $\epsilon(f)$ is integral over $k[\epsilon(g)]$. Therefore, there exist

$$d \geq 1 \quad \text{and} \quad \pi = \pi(f, g) \in f^d + \sum_{i=0}^{d-1} k[g] f^i \tag{2.2.4}$$

such that $\epsilon(\pi) = \pi(\epsilon(f), \epsilon(g)) = 0$, i.e., π belongs to $\ker \epsilon$. We note that

$$\pi, f, g \in R = M \cap k[\boldsymbol{x}] \qquad \subset \qquad k[\boldsymbol{x}]$$

$$\downarrow \epsilon \qquad\qquad\qquad\qquad \downarrow \epsilon$$

$$\epsilon(f), \epsilon(g) \in R^\epsilon \subsetneq Q(R^\epsilon) \cap k[x_1] \subset k[x_1]$$

Since $\epsilon(h) = h$ and $h = \epsilon(f)/\epsilon(g)$, we have $\epsilon(f - gh) = \epsilon(f) - \epsilon(g)h = 0$. Hence, $f - gh$ also belongs to $\ker \epsilon$. Therefore, by ($2°$), we can find $\mathbf{t} \in \mathbb{Z}^{n-1}$ such that

$$\theta(f - gh) \in k[\boldsymbol{x}] \quad \text{and} \quad \theta(\pi) = \pi(\theta(f), \theta(g)) \in x_1 k[\boldsymbol{x}]. \tag{2.2.5}$$

With this notation, the following theorem holds (cf. [76], Theorem 1.4).

Theorem 2.2.4 (Kuroda) *Let M be a subfield of $k(\boldsymbol{x})$ containing k and satisfying (A) and (B). For a polynomial h as in (2.2.2) and $\mathbf{t} \in \mathbb{Z}^{n-1}$, we define $\theta = \theta_{\mathbf{t}}^h \in \mathrm{Aut}_k k[\boldsymbol{x}, z]_{x_1}$ as in (2.2.3). If (2.2.5) holds for some $f, g \in M \cap k[\boldsymbol{x}]$ with $\epsilon(f)/\epsilon(g) = h$, and some π as in (2.2.4) with $\epsilon(\pi) = 0$, then the k-algebra $M(z)^\theta \cap k[\boldsymbol{x}, z]$ is not finitely generated.*

2.2.3 Example and Remarks

Let us illustrate how Theorem 2.2.4 can be used in the case of Example 2.2.2.

Example 2.2.2 (Continued) Since $R^\epsilon = k[x_1^2, x_1^3]$, we have

$$x_1 \in k[x_1] \setminus k[x_1^2, x_1^3] = Q(R^\epsilon) \cap k[x_1] \setminus R^\epsilon.$$

So, we take $h := x_1$. Let us give a sufficient condition on $t_2 \in \mathbb{Z}$ for $M(z)^\theta$ to be a counterexample to Hilbert's fourteenth problem.

Set

$$f := s + t = x_1 + y + x_1 y = x_1^3 + x_1 x_2 + x_2$$

$$g := s = x_1 + y = x_1^2 + x_2$$

$$\pi := f^2 - g^3.$$

Then, since $\epsilon(f) = x_1^3$ and $\epsilon(g) = x_1^2$, we have $\epsilon(f)/\epsilon(g) = x_1 = h$ and $\epsilon(\pi) = \epsilon(f)^2 - \epsilon(g)^3 = 0$.

Now, observe that $\theta(f) \in x_1^{-3} + x_1^{t_2-1} k[x]$ and $\theta(g) \in x_1^{-2} + x_1^{t_2} k[x]$. From this, it easily follows that

$$\theta(f - gh) = \theta(f) - \theta(g)x_1^{-1} \in x_1^{t_2-1} k[x]$$

$$\theta(\pi) = \theta(f)^2 - \theta(g)^3 \in x_1^{t_2-4} k[x].$$

Thus, (2.2.5) holds if $t_2 \geq 5$. Therefore, we know by Theorem 2.2.4 that

$$M(z)^\theta = k(s, t, z)^\theta = k(x_1^{t_2} x_2 + x_1^{-2}, x_1^{t_2-1} x_2 - x_1^{-2} + x_1^{-3}, z + x_1^{-1})$$

is a counterexample to Hilbert's fourteenth problem if $t_2 \geq 5$.

Remark 2.2.5

(i) $M(z) = k(x_1 + y, x_1 y, z)$ in Example 2.2.2 is equal to the invariant field $k(x_1, x_2, z)^{\langle \sigma \rangle}$, where $\sigma \in \mathrm{Aut}_k k(x_1, x_2, z)$ is such that $\sigma(x_1) = y$, $\sigma(y) = x_1$, and $\sigma(z) = z$. Accordingly, $M(z)^\theta$ is the invariant field of the subgroup $\langle \theta \circ \sigma \circ \theta^{-1} \rangle$ of $\mathrm{Aut}_k k(x_1, x_2, z)$ of order two.

(ii) More generally, if M in Theorem 2.2.4 has the form $M = k(x)^G$ for some subgroup G of $\mathrm{Aut}_k k(x)$, then $M(z)^\theta$ is the invariant field of some subgroup of $\mathrm{Aut}_k k(x, z)$ isomorphic to G. In fact, if we extend each $\sigma \in G$ to an element of $\mathrm{Aut}_k k(x, z)$ by setting $\sigma(z) = z$, and regard G as a subgroup of $\mathrm{Aut}_k k(x, z)$, then $k(x)^G(z)$ is equal to $k(x, z)^G$ (Exercise 45). Hence, we have

$$M(z)^\theta = (k(x)^G(z))^\theta = (k(x, z)^G)^\theta = k(x, z)^{\theta \circ G \circ \theta^{-1}}.$$

Some readers may find Remark 2.2.5 (i) strange, because the invariant ring $k[x, z]^H$ of every finite subgroup H of $\mathrm{Aut}_k k[x, z]$ is finitely generated by a well-known theorem of Noether [96] (Exercise 42). However, Remark 2.2.5 (i) does not say that $M(z)^\theta \cap k[x, z]$ is the invariant ring of a finite subgroup of $\mathrm{Aut}_k\, k[\boldsymbol{x}, z]$. In fact, $\theta \circ \sigma \circ \theta^{-1}$ does not restrict to an element of $\mathrm{Aut}_k\, k[\boldsymbol{x}, z]$, since $(\theta \circ \sigma \circ \theta^{-1})(x_1) = \theta(y^{-1})$ does not belong to $k[\boldsymbol{x}, z]$.

Before closing this section, let us mention an interesting consequence of Theorem 2.2.4. To state the result, we introduce the notion of minimality for subfields of $k(\boldsymbol{x})$ containing k.

Remark 2.2.6 Let L be a subfield of $k(\boldsymbol{x})$ containing k, and $A := L \cap k[\boldsymbol{x}]$.

(i) We have $Q(A) \cap k[\boldsymbol{x}] = A$, and $Q(A)$ is the least element of

$$\{L' \mid L' \text{ is a subfield of } k(\boldsymbol{x}) \text{ such that } L' \cap k[\boldsymbol{x}] = A\}$$

with respect to inclusion.

(ii) We have

$$Q(A) = L \iff \mathrm{tr.deg}_k\, A = \mathrm{tr.deg}_k\, L \qquad (2.2.6)$$

(Exercise 36 (3)). If these equivalent conditions are satisfied, then we say that L is **minimal**.

(iii) In view of (i), we may assume that L is minimal in Hilbert's fourteenth problem (see Exercise 49 for non-minimal counterexamples).

(iv) $k(S)$ is minimal for any $S \subset k[\boldsymbol{x}]$. Indeed, since $k(S) \cap k[\boldsymbol{x}]$ contains $k[S]$, we have $\mathrm{tr.deg}_k\,(k(S) \cap k[\boldsymbol{x}]) = \mathrm{tr.deg}_k\, k[S] = \mathrm{tr.deg}_k\, k(S)$.

(v) Using a result in Sect. 2.5, we can prove that, if M in Theorem 2.2.4 is minimal, then $M(z)^\theta$ is also minimal (Exercise 46 (2)).

The following surprising result is derived from Theorem 2.2.4 ([76], Theorem 1.3).

Theorem 2.2.7 (Kuroda) *Assume that M is minimal, $M \neq k(\boldsymbol{x})$, and $\mathrm{tr.deg}_k\, M \geq 2$. Then, there exists $\phi \in \mathrm{Aut}_k\, k(\boldsymbol{x}, z)$ such that $M(z)^\phi$ is minimal and a counterexample to Hilbert's fourteenth problem.*

For example, $M = k(x_1^2, x_2)$ satisfies the assumption of Theorem 2.2.7. Therefore, there exists $\phi \in \mathrm{Aut}_k\, k(x_1, x_2, z)$ for which $k(x_1^2, x_2, z)^\phi$ is minimal and a counterexample to Hilbert's fourteenth problem.

The rest of this chapter is devoted to the proof of Theorem 2.2.4 (Sects. 2.3, 2.4, and 2.5), and its applications (Sects. 2.6 and 2.7).

2.3   Criterion for Non-finite Generation

We prove the non-finite generation of $M(z)^\theta \cap k[\boldsymbol{x}, z]$ by means of the following theorem.

Theorem 2.3.1 (Kuroda) *A k-subalgebra $\mathscr{A}$ of $k[\boldsymbol{x}, z]$ is not finitely generated if the following conditions hold*:

(I) *$\mathscr{A}$ is contained in $\mathfrak{n} + k$, where $\mathfrak{n} := \sum_{i=1}^{n} x_i k[\boldsymbol{x}, z]$.*
(II) *There exists $w \in k[\boldsymbol{x}] \setminus \{0\}$ which satisfies the following condition*:
  For each $l \geq 1$, there exists $q_l \in \mathscr{A}$ of the form $q_l = \sum_{i \geq 0} q_{l,i} z^i$, where $q_{l,i} \in k[\boldsymbol{x}]$ for all i, and $q_{l,l} = c_l w$ for some $c_l \in k^*$.

Proof Set $\mathfrak{m} := \sum_{i=1}^{n} x_i k[\boldsymbol{x}]$. For each $t \geq 1$, let $\pi_t : k[\boldsymbol{x}, z] \to (k[\boldsymbol{x}]/\mathfrak{m}^t)[z]$ be the natural surjection. First, we note the following:

(i) For any $g_1, \ldots, g_t \in \mathfrak{n}$, we have $\pi_t(g_1 \cdots g_t) = 0$.
(ii) Since $w \neq 0$, there exists $e \geq 1$ such that $w \notin \mathfrak{m}^e$.
(iii) For each $p \in \mathscr{A}$, there exists $c \in k$ such that $p - c \in \mathfrak{n}$ by (I).

Now, suppose that $\mathscr{A} = k[f_1, \ldots, f_u]$ holds for some $f_1, \ldots, f_u \in \mathscr{A}$, where $u \geq 1$. By (iii), we may take $f_1, \ldots, f_u$ from $\mathfrak{n}$. Let d be an integer greater than the z-degrees of $\pi_e(f_1), \ldots, \pi_e(f_u)$, and take $q_l \in \mathscr{A}$ as in (II) for $l := de$. Then, since $q_{l,l} = c_l w$ and $w \notin \mathfrak{m}^e$, we have $\deg_z \pi_e(q_l) \geq l$. On the other hand, since q_l is in $\mathscr{A} = k[f_1, \ldots, f_u]$, we can write q_l as a k-linear combination of $f_1^{i_1} \cdots f_u^{i_u}$ for $i_1, \ldots, i_u \geq 0$. By (i), we have $\pi_e(f_1^{i_1} \cdots f_u^{i_u}) = 0$ if $i_1 + \cdots + i_u \geq e$. Thus, $\pi_e(q_l)$ is a k-linear combination of $\pi_e(f_1^{i_1} \cdots f_u^{i_u})$ for $i_1, \ldots, i_u \geq 0$ with $i_1 + \cdots + i_u < e$. When $i_1 + \cdots + i_u < e$, we have $\deg_z \pi_e(f_1^{i_1} \cdots f_u^{i_u}) < de = l$, since $\deg_z \pi_e(f_i) < d$ for $i = 1, \ldots, u$. Therefore, $\deg_z \pi_e(q_l)$ must be less than l. This is a contradiction. $\square$

2.4   Proof of Theorem 2.2.4, Part I

Our goal is to show that the k-algebra

$$\mathscr{A} := M(z)^\theta \cap k[\boldsymbol{x}, z]$$

in Theorem 2.2.4 is not finitely generated. For this purpose, it suffices to verify (I) and (II) in Theorem 2.3.1. In this section, we prove (I).

Recall that $R := M \cap k[\boldsymbol{x}] = M \cap k[\boldsymbol{x}]_{x_1}$ by (A). We also note that $(k[\boldsymbol{x}, z]_{x_1})^\theta = k[\boldsymbol{x}, z]_{x_1}$, since θ is an element of $\mathrm{Aut}_k k[\boldsymbol{x}, z]_{x_1}$. We use these equalities to prove the following lemma.

Lemma 2.4.1 *We have* $\mathscr{A} = R[z]^\theta \cap k[\boldsymbol{x}, z]$.

Proof First, we note the following:

(a) If $p, q \in M[z] \setminus \{0\}$ satisfy $p/q \in k[\boldsymbol{x}, z]_{x_1} \subset k(\boldsymbol{x})[z]$, then q divides p in $M[z]$.
    Hence, we have $M(z) \cap k[\boldsymbol{x}, z]_{x_1} = M[z] \cap k[\boldsymbol{x}, z]_{x_1}$.
(b) $M[z] \cap k[\boldsymbol{x}, z]_{x_1} = M[z] \cap k[\boldsymbol{x}]_{x_1}[z] = (M \cap k[\boldsymbol{x}]_{x_1})[z] = R[z]$ by (A).
(c) From (a) and (b), we obtain $M(z) \cap k[\boldsymbol{x}, z]_{x_1} = R[z]$.
(d) $\mathscr{A} = M(z)^\theta \cap k[\boldsymbol{x}, z] = M(z)^\theta \cap k[\boldsymbol{x}, z]_{x_1} \cap k[\boldsymbol{x}, z]$, since $k[\boldsymbol{x}, z] \subset k[\boldsymbol{x}, z]_{x_1}$.

Now, since $k[\boldsymbol{x}, z]_{x_1} = (k[\boldsymbol{x}, z]_{x_1})^\theta$ as mentioned, we have

$$M(z)^\theta \cap k[\boldsymbol{x}, z]_{x_1} = M(z)^\theta \cap (k[\boldsymbol{x}, z]_{x_1})^\theta = \left(M(z) \cap k[\boldsymbol{x}, z]_{x_1}\right)^\theta = R[z]^\theta$$

by (c). By (d), it follows that $\mathscr{A} = R[z]^\theta \cap k[\boldsymbol{x}, z]$.   □

Next, let $\tilde{\epsilon} : k[\boldsymbol{x}, z]_{x_1} \to k[x_1, z]_{x_1}$ be the substitution map defined by $x_2, \ldots, x_n \mapsto 0$. This is an extension of ϵ.

Remark 2.4.2 Assume that $p \in k[\boldsymbol{x}, z]$ satisfies $\tilde{\epsilon}(p) \in k$. Then, $p - \tilde{\epsilon}(p)$ is killed by $\tilde{\epsilon}$. This implies that $p - \tilde{\epsilon}(p)$ belongs to $\sum_{i=2}^n x_i k[\boldsymbol{x}, z]$. Thus, p belongs to $\sum_{i=2}^n x_i k[\boldsymbol{x}, z] + k \subset \mathfrak{n} + k$. Therefore, to prove (I), it suffices to show that $\mathscr{A}^{\tilde{\epsilon}}$ is contained in k.

Observe by (2.2.3) that

$$\tilde{\epsilon}(\theta(x_1)) = \tilde{\epsilon}(x_1^{-1}) = x_1^{-1} = \theta(\epsilon(x_1))$$
$$\tilde{\epsilon}(\theta(x_i)) = \tilde{\epsilon}(x_1^{t_i} x_i) = 0 = \theta(\epsilon(x_i)) \quad \text{for} \quad i = 2, \ldots, n.$$

Hence, we know that

$$R' := (R^\theta)^{\tilde{\epsilon}} = (R^\epsilon)^\theta \subset k[x_1^{-1}]. \tag{2.4.1}$$

Since z and $h(x_1^{-1})$ are invariant under $\tilde{\epsilon}$, we see that

$$\tilde{\epsilon}(\theta(z)) = \tilde{\epsilon}(z + h(x_1^{-1})) = z + h(x_1^{-1}) = z + \theta(h) \in k[x_1^{-1}, z]. \tag{2.4.2}$$

Thus, thanks to Lemma 2.4.1, we get

$$\mathscr{A}^{\tilde{\epsilon}} \subset (R[z]^\theta)^{\tilde{\epsilon}} \cap k[\boldsymbol{x}, z]^{\tilde{\epsilon}} = (R^\theta)^{\tilde{\epsilon}} [\tilde{\epsilon}(\theta(z))] \cap k[x_1, z] = R'[z + \theta(h)] \cap k[x_1, z].$$

Moreover, since $R'[z + \theta(h)] \subset k[x_1^{-1}, z]$ by (2.4.1) and (2.4.2), it follows that

$$\mathscr{A}^{\tilde{\epsilon}} \subset R'[z + \theta(h)] \cap k[x_1^{-1}, z] \cap k[x_1, z] = R'[z + \theta(h)] \cap k[z] =: B.$$

For example, we have $B = k[x_1^{-2}, x_1^{-3}, z + x_1^{-1}] \cap k[z]$ in the case of Example 2.2.2.

Thanks to Remark 2.4.2, the following lemma completes the proof of (I).

Lemma 2.4.3 B *is contained in* k.

Proof Suppose that $B \setminus k \neq \emptyset$, and take $p \in B \setminus k$. Then, since p lies in $R'[z + \theta(h)] \setminus k$, we can write

$$p = p_0(z + \theta(h))^m + p_1(z + \theta(h))^{m-1} + \cdots + p_m$$

$$= p_0 z^m + (mp_0\theta(h) + p_1)z^{m-1} + (\text{terms of lower degree in } z),$$

where $m \geq 0$, and $p_0, \ldots, p_m \in R'$ with $p_0 \neq 0$. Since $p \in B \setminus k \subset k[z] \setminus k$, we know that $m \geq 1$, $p_0 \in k^*$, and $c := mp_0\theta(h) + p_1 \in k$. Thus, we have $\theta(h) = (mp_0)^{-1}(c - p_1)$. Since p_1 is taken from R', it follows that $\theta(h)$ belongs to $R' = (R^\epsilon)^\theta$. Therefore, h belongs to R^ϵ. This contradicts the choice of h. $\square$

2.5   Proof of Theorem 2.2.4, Part II

We continue the proof of Theorem 2.2.4. In this section, we prove (II).

We begin with the following lemma. Here, we recall that $f, g \in R = M \cap k[\boldsymbol{x}]$ are such that $\epsilon(f)/\epsilon(g) = h \in k[x_1] \setminus R^\epsilon$.

Lemma 2.5.1 f *and* g *are algebraically independent over* k.

Proof Suppose that the lemma is false. Then, we have tr.deg$_k k(f, g) \leq 1$. This implies that $k(f, g) \cap k[\boldsymbol{x}] = k[p]$ for some $p \in k[\boldsymbol{x}]$ (Remark 2.2.3 (i)). Note that

$$f, g \in k(f, g) \cap k[\boldsymbol{x}] = k[p] \quad \text{and} \quad p \in k(f, g) \cap k[\boldsymbol{x}] \subset M \cap k[\boldsymbol{x}] = R.$$

Hence, $q := \epsilon(p)$ belongs to R^ϵ, and $\epsilon(f)$ and $\epsilon(g)$ belong to $\epsilon(k[p]) = k[q]$. Thus, we can write

$$h = \frac{\epsilon(f)}{\epsilon(g)} = \lambda(q) + \frac{\mu(q)}{\nu(q)}, \tag{2.5.1}$$

where $\lambda(z), \mu(z), \nu(z) \in k[z]$ are such that $\deg \mu(z) < \deg \nu(z)$. Since $q \in R^\epsilon \subset k[x_1]$ and $\mu(q)/\nu(q) = h - \lambda(q) \in k[x_1]$, we know that $\mu(z) = 0$ or $q \in k$. In either case, the right-hand side of (2.5.1) belongs to $k[q] \subset R^\epsilon$. This contradicts $h \notin R^\epsilon$. □

Remark 2.5.2

(i) By (2.2.4), (2.2.5), and Lemma 2.5.1, we know that $\theta(\pi)$ is a *nonzero* element of $R^\theta \cap x_1 k[x]$.
(ii) Since $\theta(f) \in k[x]^\theta \subset k[x]_{x_1}$, we can find $e \geq 1$ such that

$$\theta(\pi)^e \theta(f)^i \in k[x] \quad \text{for} \quad i = 0, \ldots, d - 1, \tag{2.5.2}$$

where $d \geq 1$ is as in (2.2.4).

Next, we describe $\mathscr{A}$ using a local ring. Let $k[x]_{(x_1)}$ be the localization of $k[x]$ at the prime ideal $x_1 k[x]$. Recall that

$$k[x]_{x_1} = \left\{ \frac{p}{x_1^l} \;\middle|\; p \in k[x], \, l \geq 0 \right\},$$

$$k[x]_{(x_1)} = \left\{ \frac{p}{q} \;\middle|\; p, q \in k[x], \, q \notin x_1 k[x] \right\}.$$

Hence, we see that $k[x]_{x_1} \cap k[x]_{(x_1)} = k[x]$. Thus, we get

$$k[x, z]_{x_1} \cap k[x]_{(x_1)}[z] = (k[x]_{x_1} \cap k[x]_{(x_1)})[z] = k[x, z].$$

Since $R[z]^\theta \subset k[x, z]^\theta \subset k[x, z]_{x_1}$, this equality and Lemma 2.4.1 imply

$$\mathscr{A} = R[z]^\theta \cap k[x, z] = R[z]^\theta \cap k[x, z]_{x_1} \cap k[x]_{(x_1)}[z] = R[z]^\theta \cap k[x]_{(x_1)}[z]. \tag{2.5.3}$$

Lemma 2.5.3

(i) *For each* $r \in k[x]_{x_1} \setminus k[x]$, *we have* $1/r \in k[x]_{(x_1)}$.
(ii) $1/\theta(g)$ *belongs to* $k[x]_{(x_1)}$.
(iii) *We define*

$$N := \sum_{i=0}^{d-1} f^i k \left[\pi, \frac{1}{g} \right]. \tag{2.5.4}$$

Then, we have $\theta(\pi^e N) \subset k[x]_{(x_1)}$, *where* $e \geq 1$ *is as in* (2.5.2).

Proof

(i) We can write $r = p/x_1^l$, where $p \in k[\mathbf{x}] \setminus x_1 k[\mathbf{x}]$ and $l \geq 1$. Then, $1/r = x_1^l/p$ belongs to $k[\mathbf{x}]_{(x_1)}$.

(ii) Since $\epsilon(g) \notin k$ as mentioned, there appears in g a monomial x_1^l with $l > 0$. By (2.2.3), this implies that x_1^{-l} appears in $\theta(g)$. Hence, $\theta(g)$ is an element of $k[\mathbf{x}]_{x_1} \setminus k[\mathbf{x}]$. Therefore, $1/\theta(g)$ belongs to $k[\mathbf{x}]_{(x_1)}$ by (i).

(iii) The assertion follows from (2.5.2) and (ii).   □

The proof of the following lemma is left to the reader (Exercise 39).

Lemma 2.5.4 *Let S be any commutative ring, and $p(z) \in S[z]$ a monic polynomial of degree $d \geq 1$. Then, we have $S[z] = \sum_{i=0}^{d-1} z^i S[p(z)]$.*

By (2.2.4), $\pi \in k[g][f]$ is a monic polynomial in f of degree d. Hence, for the $k[\pi, 1/g]$-module N defined in (2.5.4), we have

$$k[f, g]_g = k\left[g, \frac{1}{g}\right][f] = \sum_{i=0}^{d-1} f^i k\left[g, \frac{1}{g}\right][\pi]$$

$$= \sum_{i=0}^{d-1} f^i \left(k[g][\pi] + k\left[\frac{1}{g}\right][\pi]\right) = \sum_{i=0}^{d-1} f^i k[g][\pi] + \sum_{i=0}^{d-1} f^i k\left[\frac{1}{g}\right][\pi] \qquad (2.5.5)$$

$$= k[g][f] + N = k[f, g] + N,$$

where the second and fifth equalities are due to Lemma 2.5.4.

The following lemma is proved by using the decomposition of $k[f, g]_g$ above.

Lemma 2.5.5 *Set $r := f/g$. Then, for each $l \geq 1$, there exist sequences $f_1, \dots, f_l \in k[f, g]$ and $g_1, \dots, g_l \in N$ such that*

$$g_1 = r + f_1$$

$$g_2 = \frac{1}{2!}r^2 + f_1 r + f_2$$

$$g_3 = \frac{1}{3!}r^3 + \frac{f_1}{2!}r^2 + f_2 r + f_3$$

$$\cdots \cdots$$

$$g_l = \frac{1}{l!}r^l + \frac{f_1}{(l-1)!}r^{l-1} + \frac{f_2}{(l-2)!}r^{l-2} + \cdots + f_{l-1}r + f_l.$$

Proof

(1) Clearly, r lies in $k[f, g]_g$. Since $k[f, g]_g = k[f, g] + N$ by (2.5.5), there exists $f_1 \in k[f, g]$ such that $g_1 := r + f_1$ belongs to N.

(2) Then, $r^2/2! + f_1 r$ lies in $k[f, g]_g$. Hence, there exists $f_2 \in k[f, g]$ such that $g_2 := r^2/2! + f_1 r + f_2$ belongs to N similarly.

(3) Then, $r^3/3! + f_1 r^2/2! + f_2 r$ lies in $k[f, g]_g$. Hence, there exists $f_3 \in k[f, g]$ such that $g_3 := r^3/3! + f_1 r^2/2! + f_2 r + f_3$ belongs to N similarly.

Iterating this process, we can construct the required sequences.   □

Now, we are ready to give a

Proof of (II) By Remark 2.5.2 (i), $\theta(\pi)$ is a nonzero element of $k[x]$. We show that the required condition holds for $w := \theta(\pi)^e$. Fix an integer $l \geq 1$, and let $f_1, \ldots, f_l \in k[f, g]$ and $g_1, \ldots, g_l \in N$ be the sequences as in Lemma 2.5.5. We define

$$P_i(z) := \theta(\pi)^e \left(\frac{1}{i!} z^i + \frac{\theta(f_1)}{(i-1)!} z^{i-1} + \frac{\theta(f_2)}{(i-2)!} z^{i-2} + \cdots + \theta(f_i) \right)$$

for $i = 0, \ldots, l$. Then, we have

$$P_i(\theta(r)) = \theta(\pi)^e \theta(g_i) = \theta(\pi^e g_i) \in \theta(\pi^e N) \subset k[x]_{(x_1)} \qquad (2.5.6)$$

for $i = 0, \ldots, l$ by Lemma 2.5.3 (iii), where $g_0 := 1$.

Now, observe that $P_l(\theta(z))$ has the form

$$P_l(\theta(z)) = P_l(z + \theta(h)) = \frac{1}{l!} \theta(\pi)^e z^l + \text{(terms of lower degree in } z\text{)}.$$

So, we show that $P_l(\theta(z))$ belongs to $\mathscr{A}$. Since $P_l(z) \in k[f, g]^\theta[z] \subset R^\theta[z]$ by definition, we have

$$P_l(\theta(z)) \in R^\theta[\theta(z)] = R[z]^\theta.$$

Thus, in view of (2.5.3), it suffices to check that $P_l(\theta(z))$ belongs to $k[x]_{(x_1)}[z]$.
By Taylor's formula, we have

$$P_l(\theta(z)) = P_l(\theta(r) + \theta(z - r)) = \sum_{i=0}^{l} \frac{P_{l-i}(\theta(r))}{i!} \theta(z - r)^i. \qquad (2.5.7)$$

By (2.2.5) and Lemma 2.5.3 (ii), we know that

$$\theta(z - r) = z + \theta(h) - \frac{\theta(f)}{\theta(g)} = z + \frac{\theta(gh - f)}{\theta(g)} \in k[\boldsymbol{x}]_{(x_1)}[z]. \tag{2.5.8}$$

From (2.5.6), (2.5.7), and (2.5.8), we conclude that $P_l(\theta(z))$ belongs to $k[\boldsymbol{x}]_{(x_1)}[z]$. $\square$

This completes the proof of Theorem 2.2.4.

2.6 Application (1): Derivations

As an application of Theorem 2.2.4, we construct k-derivations of $k[\boldsymbol{x}, z]$ whose kernels are not finitely generated.

Remark 2.6.1 If M in Theorem 2.2.4 has the form $M = k(\boldsymbol{x})^D$ for some k-derivation D of $k(\boldsymbol{x})$, then $M(z)^\theta \cap k[\boldsymbol{x}, z]$ is equal to the kernel of the k-derivation $\widetilde{D}$ of $k[\boldsymbol{x}, z]$ defined as follows:

(1) Extend D to a k-derivation of $k(\boldsymbol{x}, z)$ by setting $D(z) = 0$. Then, we have $k(\boldsymbol{x}, z)^D = k(\boldsymbol{x})^D(z)$ (Exercise 44).
(2) Define a k-derivation D' of $k(\boldsymbol{x}, z)$ by $D' := \theta \circ D \circ \theta^{-1}$. Then, we have

$$M(z)^\theta = (k(\boldsymbol{x})^D(z))^\theta = (k(\boldsymbol{x}, z)^D)^\theta = k(\boldsymbol{x}, z)^{D'}.$$

(3) Since $D'(x_1), \ldots, D'(x_n)$ and $D'(z)$ are elements of $k(\boldsymbol{x}, z)$, we can find $p \in k(\boldsymbol{x}, z)^*$ such that $pD'(x_1), \ldots, pD'(x_n)$ and $pD'(z)$ belong to $k[\boldsymbol{x}, z]$. Then, $\widetilde{D} := pD'$ restricts to a k-derivation of $k[\boldsymbol{x}, z]$. Moreover, we have

$$M(z)^\theta \cap k[\boldsymbol{x}, z] = k(\boldsymbol{x}, z)^{D'} \cap k[\boldsymbol{x}, z] = k(\boldsymbol{x}, z)^{\widetilde{D}} \cap k[\boldsymbol{x}, z] = k[\boldsymbol{x}, z]^{\widetilde{D}}.$$

In view of Remark 2.6.1, it suffices to construct a k-derivation D of $k(\boldsymbol{x})$ for which $M := k(\boldsymbol{x})^D$ satisfies (A) and (B). Take any

$$p_2 \in k[x_1], \; p_3 \in k[x_1, x_2], \ldots, p_n \in k[x_1, \ldots, x_{n-1}],$$

and set

$$y_1 := x_1, \quad \text{and} \quad y_i := x_i + p_i \quad \text{for} \quad i = 2, \ldots, n.$$

Then, $y_1, \ldots, y_n$ form a system of variables of $k[x]$. We define

$$D := \frac{\partial}{\partial y_1} \quad \text{and} \quad M := k(x)^D = k(y_2, \ldots, y_n).$$

In the notation above, the following theorem holds.

Theorem 2.6.2 *Assume that there does not exist $p \in k[x_1]$ satisfying*

$$k[\epsilon(p_2), \ldots, \epsilon(p_n)] = k[p].$$

Then, M satisfies (A) *and* (B).

Proof Since $y_1, \ldots, y_n$ form a system of variables of $k[x]$, we see that

$$R = M \cap k[x] = k(y_2, \ldots, y_n) \cap k[y_1, \ldots, y_n] = k[y_2, \ldots, y_n],$$

$$M \cap k[x]_{x_1} = k(y_2, \ldots, y_n) \cap k[y_1, y_1^{-1}, y_2, \ldots, y_n] = k[y_2, \ldots, y_n].$$

Hence, (A) holds true. Since $R^\epsilon = k[y_2, \ldots, y_n]^\epsilon = k[\epsilon(p_2), \ldots, \epsilon(p_n)]$, we have (B) by assumption.   □

Let us give an example to illustrate Theorem 2.6.2.

Example 2.6.3 Assume that $n = 3$, and let $p_2 = x_1^2$ and $p_3 = x_1^3$. Then, the assumption of Theorem 2.6.2 is satisfied. Hence, (A) and (B) hold for $M := k(y_2, y_3)$. In this case, we can take

$$f := y_3 = x_3 + x_1^3, \quad g := y_2 = x_2 + x_1^2, \quad h := x_1 \quad \text{and} \quad \pi := f^2 - g^3.$$

Note that $\theta(f) \in x_1^{-3} + x_1^{t_3} k[x]$ and $\theta(g) \in x_1^{-2} + x_1^{t_2} k[x]$. Hence, if $t_2 \geq 5$ and $t_3 \geq 4$, then we have

$$\theta(f - gh) = \theta(f) - \theta(g)x_1^{-1} \in x_1^4 k[x] \quad \text{and} \quad \theta(\pi) = \theta(f)^2 - \theta(g)^3 \in x_1 k[x].$$

Thus, (2.2.5) is satisfied.
    Now, write

$$D := \frac{\partial}{\partial y_1} = \frac{\partial}{\partial x_1} - 2x_1 \frac{\partial}{\partial x_2} - 3x_1^2 \frac{\partial}{\partial x_3}.$$

Regard this D as a k-derivation of $k(x, z)$ with $D(z) = 0$, and set $D' := \theta \circ D \circ \theta^{-1}$. Then, we can write

$$D' = -x_1^2 \frac{\partial}{\partial x_1} + (t_2 x_1 x_2 - 2x_1^{-t_2-1}) \frac{\partial}{\partial x_2} + (t_3 x_1 x_3 - 3x_1^{-t_3-2}) \frac{\partial}{\partial x_2} - \frac{\partial}{\partial z}.$$

So, take an integer $u \geq \max\{t_2 + 1, t_3 + 2\}$ and define $\widetilde{D} := x_1^u D'$. Then, $\widetilde{D}$ restricts to a k-derivation of $k[x, z]$. By construction, $k[x, z]^{\widetilde{D}}$ is equal to $M(z)^\theta \cap k[x, z]$, which is not finitely generated if $t_2 \geq 5$ and $t_3 \geq 4$.

The following lemma is useful to check (A) in more general cases.

Lemma 2.6.4 Let D be a k-derivation of $k[x]$ with $D(x_1) \notin x_1 k[x]$. Then, we have $k(x)^D \cap k[x]_{x_1} = k(x)^D \cap k[x]$.

Proof If the assertion is false, then there exists $q \in k[x]_{x_1} \setminus k[x]$ such that $D(q) = 0$. Write $q = p/x_1^l$, where $p \in k[x] \setminus x_1 k[x]$ and $l \geq 1$. Then, we have

$$0 = D\left(\frac{p}{x_1^l}\right) = \frac{D(p)x_1^l - p \cdot lx_1^{l-1}D(x_1)}{x_1^{2l}} = \frac{D(p)x_1 - lpD(x_1)}{x_1^{l+1}}.$$

Hence, we get $D(p)x_1 = lpD(x_1)$. Thus, $pD(x_1)$ belongs to $x_1 k[x]$. Therefore, p or $D(x_1)$ belongs to $x_1 k[x]$, a contradiction.   □

Remark 2.6.5 Assume that D is a locally nilpotent k-derivation of $k[x]$ with $D(x_1) \neq 0$. Then, $D(x_1)$ does not belong to $x_1 k[x]$ by the *eigenvalue property* (cf. [117], Proposition 1.3.32 ii)). Hence, we always have $k(x)^D \cap k[x]_{x_1} = k(x)^D \cap k[x]$ by Lemma 2.6.4.

2.7   Application (2): Invariant Fields

In this section, we generalize Example 2.2.2. Assume that $n \geq 2$, and set

$$y_1 := x_1, \quad y_2 := x_2 - x_1 + x_1^2, \quad \text{and} \quad y_i := x_i \quad \text{for} \quad i = 3, \dots, n.$$

Then, $y_1, \dots, y_n$ form a system of variables of $k[x]$. Let G be a permutation group on the set $\{y_1, \dots, y_n\}$ with the following condition:

There exists $2 \leq l \leq n$ such that $\{y_1, \dots, y_l\}$ is the G-orbit of y_1.   (2.7.1)

We identify each $\sigma \in G$ with the element of $\mathrm{Aut}_k\, k[x]$ defined by $y_i \mapsto \sigma(y_i)$ for $i = 1, \ldots, n$. Then, we can regard G as a subgroup of $\mathrm{Aut}_k\, k[x]$.

Example 2.7.1 Let ρ be the cyclic permutation on $\{y_1, \ldots, y_n\}$ defined by $y_1 \mapsto y_2 \mapsto \cdots \mapsto y_l \mapsto y_1$ and $y_i \mapsto y_i$ for $i = l+1, \ldots, n$. Then, $G := \langle \rho \rangle$ satisfies (2.7.1).

Remark 2.7.2 By (2.7.1), G induces a permutation group on $\{y_1, \ldots, y_l\}$. Since symmetric polynomials in $y_1, \ldots, y_l$ are invariant under *arbitrary* permutations on $\{y_1, \ldots, y_l\}$, the invariant ring $k[x]^G$ contains all symmetric polynomials in $y_1, \ldots, y_l$.

The following theorem is a generalization of Example 2.2.2 (cf. [76], Theorem 4.1).

Theorem 2.7.3 *Let G be as above, and let $\theta = \theta_{\mathbf{t}}^h \in \mathrm{Aut}_k\, k[x, z]_{x_1}$ be as in (2.2.3) with $h = x_1$ and $\mathbf{t} \in \mathbb{Z}^{n-1}$. If $t_2 \geq 5$ and $t_i \geq 6$ for $i = 3, \ldots, l$, then $(k(x)^G(z))^\theta$ is a counterexample to Hilbert's fourteenth problem.*

Proof First, we show that (A) and (B) hold for $M = k(x)^G$. By (2.7.1), there exists $\tau \in G$ such that $\tau(y_1) = y_2$. Since $y_1 = x_1$ by definition, we have

$$k(x)^G \cap k[x]_{x_1} = k(x)^G \cap k[x]_{y_2} = k(x)^G \cap k[x]_{x_1} \cap k[x]_{y_2} = k(x)^G \cap k[x]$$

similarly to (2.2.1). Hence, (A) holds true.

For (B), note that $M \cap k[x] = k[x]^G$. We show that $(k[x]^G)^\epsilon = k[x_1^2, x_1^3]$. Write $p \in k[x]^G$ as

$$p = c + \sum_{i=1}^{n} c_i y_i + (\text{monomials of degree} \geq 2 \text{ in } y_1, \ldots, y_n),$$

where $c, c_1, \ldots, c_n \in k$. Then, c_1 must be equal to c_2, since $\tau(y_1) = y_2$ and $\tau(p) = p$. On the other hand, we have

$$\epsilon(y_1) = x_1, \quad \epsilon(y_2) = -x_1 + x_1^2 \quad \text{and} \quad \epsilon(y_i) = 0 \quad \text{for} \quad i = 3, \ldots, n.$$

Thus, we know that $\epsilon(p)$ has no linear term, i.e.,

$$\epsilon(p) \in k + kx_1^2 + kx_1^3 + kx_1^4 + \cdots = k[x_1^2, x_1^3].$$

Therefore, $(k[x]^G)^\epsilon$ is contained in $k[x_1^2, x_1^3]$. Next, we define

$$g := y_1 + \cdots + y_l = x_1^2 + \underline{x_2} + y,$$

$$f := \sum_{1 \le i < j \le l} y_i y_j + g = (x_1(y_2 + y) + y_2 y + y') + x_1^2 + x_2 + y$$

$$= x_1 y_2 + x_1^2 + (x_1 + y_2)y + y' + x_2 + y$$

$$= x_1^3 + x_1 x_2 + (x_2 + x_1^2)y + y' + x_2 + y,$$

where

$$y := \sum_{i=3}^{l} y_i \quad \text{and} \quad y' := \sum_{3 \le i < j \le l} y_i y_j.$$

Since f and g are symmetric polynomials in $y_1, \ldots, y_l$, we have $f, g \in k[x]^G$ by Remark 2.7.2. Since $\epsilon(f) = x_1^3$ and $\epsilon(g) = x_1^2$, it follows that x_1^3 and x_1^2 belong to $(k[x]^G)^\epsilon$, proving $(k[x]^G)^\epsilon = k[x_1^2, x_1^3]$. Therefore, (B) holds true.

Next, note that $\epsilon(f)/\epsilon(g) = x_1$ and $\pi := f^2 - g^3 \in \ker \epsilon$. Note also that $\theta(g) \in x_1^{-2} + x_1^5 k[x]$ and $\theta(f) \in x_1^{-3} + x_1^4 k[x]$, since $t_2 \ge 5$ and $t_i \ge 6$ for $i = 3, \ldots, l$ by assumption. Hence, $\theta(f - gx_1) = \theta(f) - \theta(g)x_1^{-1}$ and $\theta(\pi) = \theta(f)^2 - \theta(g)^3$ belong to $x_1^4 k[x]$ and $x_1 k[x]$, respectively. Thus, (2.2.5) is satisfied. Therefore, the theorem follows from Theorem 2.2.4.   □

As discussed in Remark 2.2.5 (ii), $(k(x)^G(z))^\theta$ in Theorem 2.7.3 is the invariant field of a subgroup of $\text{Aut}_k k(x, z)$ isomorphic to G. It is interesting to mention that the invariant field of this type may fail to be a purely transcendental extension of k when $k = \mathbb{Q}$ (Exercise 48).

Exercises for This Chapter Let k be any field unless otherwise stated, and let z and T be variables.

Exercise 36 Let B be an integrally closed domain, L a subfield of $Q(B)$, and $A := L \cap B$.

(1) Show that A is integrally closed in L (Since $Q(A) \subset L$, this implies that A is integrally closed in $Q(A)$). [If $b \in L \subset Q(B)$ is integral over $A \subset B$, then b is integral over B, and so $b \in B \cap L = A$ by the assumption that B is integrally closed.]

(2) Deduce from (1) that $Q(A)$ is algebraically closed in L. [Assume that $a_0 + a_1 b + \cdots + a_l b^l = 0$, $a_l \ne 0$ holds with $b \in L$ and $a_i \in A$. Then, $a_l b$ is integral over A, since $a_0 a_l^{l-1} + a_1 a_l^{l-2}(a_l b) + \cdots + (a_l b)^l = 0$. Hence, we have $a_l b \in A$ by (1), and so $b \in Q(A)$.]

(3) Deduce (2.2.6) from (2). [Recall that $k[x]$ is an integrally closed domain.]

Exercise 37 Prove Remark 2.2.3 (i). [Thanks to Zaks' theorem ([133], or [117], Theorem 1.2.26 and the notes at the end of Chapter 1), it suffices to check that $L \cap k[x]$ is equal to k or a **Dedekind ring**, i.e., integrally closed (Exercise 36 (1)), Noetherian, and of Krull dimension one.]

Exercise 38 Let A be a k-subalgebra of $k[z]$ with $A \neq k$.

(1) Show that $k[z]$ is integral over A (or equivalently, z is integral over A). [If $p(z) \in A \setminus k$ is monic, then $\phi(T) := p(T) - p(z) \in A[T]$ is monic and satisfies $\phi(z) = 0$.]
(2) Show that $Q(A) \cap k[z]$ is the integral closure of A in $Q(A)$. [$k[z]$ is integrally closed.]
(3) Show that A is integrally closed if and only if there exists $f \in A$ such that $A = k[f]$. [If A is integrally closed, then $A = Q(A) \cap k[z]$ by (2). This implies that $A = k[f]$ for some $f \in A$ by Remark 2.2.3 (i).]

Exercise 39 Prove Lemma 2.5.4.
[Note that $\phi(T) := p(T) - p(z)$ is a monic polynomial in T of degree d over $S[p(z)]$. Take any $q(z) \in S[z]$, and let $r(T) \in \sum_{i=0}^{d-1} T^i S[p(z)]$ be the remainder of $q(T) \in S[T] \subset S[p(z)][T]$ divided by $\phi(T)$. Then, we obtain $q(z) = r(z) \in \sum_{i=0}^{d-1} z^i S[p(z)]$ by the substitution $T \mapsto z$.]

Exercise 40 Let R be a Noetherian ring, $B = R[b_1, \ldots, b_n]$ a finitely generated R-algebra, and A an R-subalgebra of B such that $b_1, \ldots, b_n$ are integral over A. Show the following:

(1) There exist finitely many $a_1, \ldots, a_m \in A$ such that $b_1, \ldots, b_n$ are integral over $R' := R[a_1, \ldots, a_m]$. [If $b \in B$ satisfies $f(b) = 0$ for a monic polynomial $f(T) = T^l + \alpha_1 T^{l-1} + \cdots + \alpha_l \in A[T]$, then b is integral over $R[\alpha_1, \ldots, \alpha_l]$.]
(2) A is finitely generated as an R'-module. [B is finitely generated as an R'-module, R' is Noetherian, and A is an R'-submodule of B.]
(3) A is finitely generated as an R-algebra.

Exercise 41 Let R and B be as in Exercise 40, and D an R-derivation of B. Show that, if char $R > 0$, then the R-algebra $B^D = \{b \in B \mid D(b) = 0\}$ is finitely generated.
[If $m := $ char R, then b_i^m belongs to B^D for $i = 1, \ldots, n$, since $D(b_i^m) = mb_i^{m-1}D(b_i) = 0$. This implies that $b_1, \ldots, b_n$ are integral over B^D.]

Exercise 42 Let R and B be as in Exercise 40, and G a finite subgroup of $\mathrm{Aut}_R B$. Show that the invariant ring B^G is finitely generated as an R-algebra. [Each $b \in B$ is integral over B^G, since b is a root of the monic polynomial $\phi(T) = \prod_{\sigma \in G}(T - \sigma(b)) \in B^G[T]$.]

In Exercises 41 and 42, we cannot drop the assumption that R is Noetherian, as the following exercise shows.

Exercise 43 Let $k[a, b]$ be the polynomial ring in two variables over k, and let $R :=$ $k + ak[a, b]$. Then, R is a k-subalgebra of $k[a, b]$, which is not Noetherian, since the ideal $ak[a, b]$ of R is not finitely generated.

(1) Let $R[x_1, x_2]$ be the polynomial ring in two variables over R. We define an R-derivation D of $R[x_1, x_2]$ by

$$D = ab \frac{\partial}{\partial x_1} - a \frac{\partial}{\partial x_2}.$$

Show that the kernel $R[x_1, x_2]^D$ is not finitely generated as an R-algebra.
[Let $R[x_1, x_2] = \bigoplus_{l=0}^\infty R[x_1, x_2]_l$ be the standard grading on $R[x_1, x_2]$, i.e., $R[x_1, x_2]_l$ is the R-module generated by monomials in x_1 and x_2 of total degree l. Then, we have

$$R[x_1, x_2]^D = \bigoplus_{l=0}^\infty (R[x_1, x_2]^D \cap R[x_1, x_2]_l),$$

since $D(R[x_1, x_2]_l) \subset R[x_1, x_2]_{l-1}$ for each $l \geq 1$. Hence, if the R-algebra $R[x_1, x_2]^D$ is finitely generated, then it is generated by finitely many homogeneous polynomials. This implies that the R-module

$$R[x_1, x_2]_1 \cap R[x_1, x_2]^D = \{r(x_1 + bx_2) \mid r \in R \cap b^{-1}R\}$$

$$\simeq R \cap b^{-1}R = ak[a, b]$$

is finitely generated, which is absurd.]
(2) Let $R[x_1, x_2, x_3]$ be the polynomial ring in three variables over R, and define $\phi \in$ $\mathrm{Aut}_R R[x_1, x_2, x_3]$ by

$$\phi(x_1) = x_1 + abx_3, \quad \phi(x_2) = x_2 - ax_3 \quad \text{and} \quad \phi(x_3) = x_3.$$

Show that the invariant ring $R[x_1, x_2, x_3]^{\langle \phi \rangle}$ is not finitely generated as an R-algebra. Show also that $\langle \phi \rangle \simeq \mathbb{Z}/p\mathbb{Z}$, where $p := \mathrm{char}\, k \geq 0$. [Use the same technique as (1).]

Exercise 44 Let G be a subgroup of $\mathrm{Aut}_k k(\mathbf{x})$. We extend each $\sigma \in G$ to an element of $\mathrm{Aut}_k k(\mathbf{x}, z)$ by setting $\sigma(z) = z$ and regard G as a subgroup of $\mathrm{Aut}_k k(\mathbf{x}, z)$.

(1) Show that $k(\mathbf{x})[z]^G = k(\mathbf{x})^G[z]$.
(2) Show that $k(\mathbf{x}, z)^G = k(\mathbf{x})^G(z)$. [We can *uniquely* write $r \in k(\mathbf{x}, z)^G$ as $r = p/q$, where $p, q \in k(\mathbf{x})[z]$ are such that $\gcd(p, q) = 1$, and q is monic in z. For each $\sigma \in G$, we have $r = \sigma(r) = \sigma(p)/\sigma(q)$, $\gcd(\sigma(p), \sigma(q)) = 1$, and $\sigma(q)$ is monic in z. Then, we get $\sigma(p) = p$ and $\sigma(q) = q$ by uniqueness.]

Exercise 45 Let D be a k-derivation of $k(x)$. We extend D to a k-derivation of $k(x, z)$ by setting $D(z) = 0$.

(1) Show that $k(x)[z]^D = k(x)^D[z]$.
(2) Show that $k(x, z)^D = k(x)^D(z)$. [Write $r \in k(x, z)^D$ as in Exercise 44 (2). Since $(D(p)q - pD(q))/q^2 = D(r) = 0$, we have $D(p)q = pD(q)$. Since q is monic in z, and $D(z) = 0$, we also have $\deg_z D(q) < \deg_z q$. Since $\gcd(p, q) = 1$, these conditions imply $D(p) = D(q) = 0$.]

Exercise 46 In the situation of Theorem 2.2.4, let $\mathscr{A} := M(z)^\theta \cap k[x]$.

(1) Show that $R[z]^\theta \subset Q(\mathscr{A})$. [By Remark 2.5.2 (i), $\theta(\pi)$ is a nonzero element of $R^\theta \cap x_1 k[x] \subset \mathscr{A}$. For each $p \in R[z]^\theta \subset k[x, z]_{x_1}$, there exists $l \geq 0$ such that $\theta(\pi)^l p \in k[x, z] \cap R[z]^\theta = \mathscr{A}$. Hence, we get $p \in \mathscr{A}[1/\theta(\pi)] \subset Q(\mathscr{A})$.]
(2) Prove Remark 2.2.6 (v) (Show that $Q(R) = M$ implies $Q(\mathscr{A}) = M(z)^\theta$).
    $[Q(R) = M$ implies $Q(R[z]^\theta) = M(z)^\theta$.]

Exercise 47 Check that the counterexamples to Hilbert's fourteenth problem given in Theorems 2.6.2 and 2.7.3 are minimal.
[By Remark 2.2.6 (iii), it suffices to check that $Q(R) = M$.]

Exercise 48 (Noether's Problem) Let G be a finite group with $|G| = n$, and write $k(x) = k(\{x_\sigma \mid \sigma \in G\})$. We define $k(G)$ to be the invariant subfield of $k(x)$ for the G-action defined by $\tau \cdot x_\sigma := x_{\tau\sigma}$ for each $\tau, \sigma \in G$. Then, **Noether's problem** (cf. [97]) asks whether $k(G)$ is **rational** over k, i.e., a purely transcendental extension of k. The following facts are known:

(1) $\mathbb{Q}(\mathbb{Z}/p\mathbb{Z})$ is not rational over $\mathbb{Q}$ for various primes p, say $p = 47$ (cf. [113]).
(2) When G is a finite abelian group, $\mathbb{Q}(G)$ is rational over $\mathbb{Q}$ if and only if $\mathbb{Q}(G)(z_1, \ldots, z_d)$ is rational over $\mathbb{Q}$ for some $d \geq 0$, where $z_1, \ldots, z_d$ are indeterminates over $\mathbb{Q}(G)$ (cf. [43]).

Deduce from (1), (2), Remark 2.2.5 (ii), and Theorem 2.7.3 that there exists a finite subgroup Γ of $\text{Aut}_\mathbb{Q} \mathbb{Q}(x, z)$ such that $\mathbb{Q}(x, z)^\Gamma$ is not rational over $\mathbb{Q}$, and is a counterexample to Hilbert's fourteenth problem. [For ρ in Example 2.7.1, $k(x)^{\langle \rho \rangle}$ is isomorphic to $k(\mathbb{Z}/l\mathbb{Z})(x_{l+1}, \ldots, x_n)$.]

Exercise 49 Assume that $\text{char } k = 0$. Let B be a k-domain, and D a nonzero locally nilpotent k-derivation of B. In (1) and (2) of this exercise, we construct $\iota \in \text{Aut}_k Q(B)$ such that $\iota^2 = \text{id}$ and $Q(B)^{\langle \iota \rangle} \cap B = B^D$ (see [76], §4.6). For simplicity, we put $L := Q(B^D)$.

(1) Find $p \in B$ such that $B \subset L[p]$, $Q(B) = L(p)$ and p is transcendental over L. [Since D is nonzero and locally nilpotent, there exists $p \in B$ with $D(p) \neq 0$ and $D^2(p) = 0$. Such a p, called a *preslice* of D, satisfies the required condition (see [117], (1.3.29) and (1.3.30)).]

(2) We define $\iota \in \mathrm{Aut}_L L(p) \subset \mathrm{Aut}_k L(p) = \mathrm{Aut}_k Q(B)$ by $\iota(p) = p^{-1}$. Show that $\iota^2 = \mathrm{id}$ and $Q(B)^{\langle \iota \rangle} \cap B = B^D$. [We have the following equalities:

$$Q(B)^{\langle \iota \rangle} = L(p)^{\langle \iota \rangle} = L(p + p^{-1}), \quad L(p + p^{-1}) \cap L[p] = L,$$

$$Q(B)^{\langle \iota \rangle} \cap B = Q(B)^{\langle \iota \rangle} \cap L[p] \cap B \quad \text{and} \quad L \cap B = B^D.]$$

(3) Let D be your favorite locally nilpotent k-derivation of $k[x]$ such that $k[x]^D$ is not finitely generated (For example, take D in [117], Theorem 9.6.2 or 9.6.17). Construct $\iota \in \mathrm{Aut}_k k(x)$ such that $\iota^2 = \mathrm{id}$ and $k(x)^{\langle \iota \rangle} \cap k[x] = k[x]^D$.

Remark $k(x)^{\langle \iota \rangle}$ constructed in (3) is a counterexample to Hilbert's fourteenth problem. However, $k(x)^{\langle \iota \rangle}$ is not minimal because tr.deg$_k k(x)^{\langle \iota \rangle} = n$ and tr.deg$_k k[x]^D = n - 1$.

Prime Characteristic Methods and the Cancellation Problem

3

3.1 The Makar-Limanov and Derksen Invariants

In studying the linearization conjecture for $\mathbb{C}^*$-actions on $\mathbb{C}^3$, Mariusz Koras and Peter Russell generated a list of hypersurfaces which could potentially provide a counterexample, the most well known being the hypersurface V in $\mathbb{C}^4$ given by

$$x + x^2 y + z^2 + t^3 = 0.$$

(See Example 9.4.14, [117]) Each of the Koras–Russell hypersurfaces admits a non-linearizable $\mathbb{C}^*$-action, and some are similar to $\mathbb{C}^3$ in many tempting ways. (The hypersurface V above is smooth, factorial, contractible, diffeomorphic to $\mathbb{R}^6$, etc.) If any hypersurface on the list turned out to be algebraically isomorphic to $\mathbb{C}^3$, then it would serve as a counterexample to the linearization conjecture. The proof of the linearization conjecture for $\mathbb{C}^*$-actions on $\mathbb{C}^3$ (Theorem 9.4.15, [117]) therefore required a technique for distinguishing each of these potential hypersurfaces from $\mathbb{C}^3$. This technique came from examining the intersection of the kernels of all locally nilpotent derivations on the coordinate ring of the hypersurface. This intersection is now known as the Makar-Limanov invariant.

Definition 3.1.1 Let R be a ring of characteristic zero. The **Makar-Limanov invariant** of R is

$$\mathrm{ML}(R) = \bigcap_{D \in \mathrm{LND}\, R} R^D.$$

If W is an algebraic variety with coordinate ring $A(W)$, define $\mathrm{ML}(W) = \mathrm{ML}(A(W))$.

© The Author(s), under exclusive license to Springer Nature Switzerland AG 2021
A. van den Essen et al., *Polynomial Automorphisms and the Jacobian Conjecture*,
Frontiers in Mathematics, https://doi.org/10.1007/978-3-030-60535-3_3

Example 3.1.2 If k is a field of characteristic zero, then $\mathrm{ML}(k^n) = \mathrm{ML}(k^{[n]}) = k$ for $n \geq 1$. (See §1.3 Exercise 11, [117])

In [77] Leonid Makar-Limanov used this invariant to show $V \not\cong \mathbb{C}^3$ by showing $\mathrm{ML}(V) \neq \mathbb{C}$. While he did not explicitly define the Makar-Limanov invariant in that paper, he did show that $x \in \mathrm{ML}(V)$, where $x \in \mathbb{C}[X, Y, Z, T]/(X + X^2Y + Z^2 + T^3)$ is the congruence class of X. More precisely, it turns out that $\mathrm{ML}(V) = \mathbb{C}[x]$ as shown in [78].

Studying Makar-Limanov's approach in [28], Harm Derksen was able to give an alternate proof that $V \not\cong \mathbb{C}^3$ by looking instead at the algebra now known as the Derksen invariant.

Definition 3.1.3 Let R be a ring of characteristic zero. The **Derksen invariant** of R, denoted $\mathscr{D}(R)$, is the subalgebra of R generated by the kernels of all nontrivial locally nilpotent derivations of R. If W is an algebraic variety with coordinate ring $A(W)$, define $\mathscr{D}(W) = \mathscr{D}(A(W))$.

Example 3.1.4 If k is a field of characteristic zero, then $\mathscr{D}(k^n) = \mathscr{D}(k^{[n]}) = k^{[n]}$ for $n \geq 2$. (See §1.3 Exercise 11, [117])

Derksen showed that $y \notin \mathscr{D}(V)$, where $y \in \mathbb{C}[X, Y, Z, T]/(X + X^2Y + Z^2 + T^3)$ is the congruence class of Y. Therefore $\mathscr{D}(V) = \mathscr{D}(A(V)) \neq A(V)$, and so $V \not\cong \mathbb{C}^3$.

With the example of k^n in mind, when studying a k-algebra A we call $\mathrm{ML}(A)$ trivial if $\mathrm{ML}(A) = k$ (as small as possible) and $\mathscr{D}(A)$ trivial if $\mathscr{D}(A) = A$ (as large as possible). One might think that both invariants give the same information, but it is possible to make examples where one is trivial while the other is not. (See Sect. 3.1 Exercise 4 and Sect. 3.3 Exercise 3.) Consequently, each invariant has found its own use. The key to successfully using either invariant is being able to compute it. For either invariant, computation often involves the use of filtrations and associated graded rings.

Definition 3.1.5 Let k be a field. Let A be a k-algebra. A $\mathbb{Z}$-**filtration** on A is a collection $\{A_n\}_{n \in \mathbb{Z}}$ of k-linear subspaces such that

1. $A_n \subseteq A_{n+1}$ for all $n \in \mathbb{Z}$,
2. $A_n A_m \subseteq A_{n+m}$ for all $n, m \in \mathbb{Z}$, and
3. $\bigcup_{n \in \mathbb{Z}} A_n = A$.

This filtration is called **proper** if it additionally satisfies

1. $\bigcap_{n \in \mathbb{Z}} A_n = \{0\}$ and
2. If $a \in A_n \setminus A_{n-1}$ and $b \in A_m \setminus A_{m-1}$, then $ab \in A_{n+m} \setminus A_{n+m-1}$ for all $a, b \in A$ and all $n, m \in \mathbb{Z}$.

Given a $\mathbb{Z}$-filtration on A, for each $a \in A$ we define the **degree** of a by $\deg(a) = \min\{n \in \mathbb{Z} \mid a \in A_n\}$. Note that deg is a degree function if the filtration is proper. From a $\mathbb{Z}$-filtration on A we construct the **associated graded ring**

$$\operatorname{gr} A = \bigoplus_{n \in \mathbb{Z}} A_n/A_{n-1}.$$

Nonzero elements of A_n/A_{n-1} are called **homogeneous of degree** n.

Note that gr A is a k-algebra, and if A is an integral domain with a proper $\mathbb{Z}$-filtration, then gr A is also an integral domain. Also note that a degree function on A determines a proper $\mathbb{Z}$-filtration. We often construct a $\mathbb{Z}$-filtration on A by assigning integer **weights** to a suitable set of generators for A to obtain a degree function. This is the most commonly used scenario.

Definition 3.1.6 Let k be a field. Let A be a k-algebra. Let $\{A_n\}$ be a $\mathbb{Z}$-filtration on A. For $D \in \operatorname{Der} A$, let

$$d = \max\{\deg(Da) - \deg(a) \mid 0 \neq a \in A\}.$$

We say a derivation $D \in \operatorname{Der} A$ **respects the filtration** if d is an integer. In this case we define the linear map gr $D : \operatorname{gr} A \to \operatorname{gr} A$ given by

$$\operatorname{gr}(D)(a + A_{n-1}) = Da + A_{n+d-1}, \text{ for all } a \in A_n.$$

If $D \in \operatorname{Der} A$ respects a given filtration on A, then gr D is a derivation on the associated graded ring gr A. (See §1.2 Exercise 7, [117]) Moreover, if D is locally nilpotent, then so is gr D. Finally, note that by the choice of d above, if $D \neq 0$, then gr $D \neq 0$. This simple observation is essential to using filtrations to compute the Makar-Limanov and Derksen invariants. We use the term **homogenization** for the process of passing from $D \in \operatorname{Der} A$ to gr $D \in \operatorname{Der}(\operatorname{gr} A)$.

Without going into detail, we mention that in order to show that the Koras–Russell hypersurface V is not isomorphic to $\mathbb{C}^3$, the arguments made by both Makar-Limanov and Derksen begin as follows. Let $A = A(V) = \mathbb{C}[X, Y, Z, T]/(X + X^2Y + Z^2 + T^3)$ and let $x, y, z, t \in A$ be the coordinate functions, that is, the congruence classes of X, Y, Z, T in A. View A as a subring of $\mathbb{C}[x, x^{-1}, z, t]$ with $y = -x^{-2}(x + z^2 + t^3)$ and assign degrees to the coordinate functions as follows: $\deg(x) = -1$, $\deg(y) = 2$, $\deg(z) = \deg(t) = 0$. From this we have a degree function which induces a $\mathbb{Z}$-filtration on A, and the associated graded ring is gr $A \cong \mathbb{C}[x, y, z, t]/(x^2y + z^2 + t^3)$. If $0 \neq D \in \operatorname{LND} A$, then gr $D \in \operatorname{LND}(\operatorname{gr} A)$ and gr $D \neq 0$. The investigation continues by considering various cases involving the kernels A^D and $\operatorname{gr}(A)^{\operatorname{gr} D}$. See [28], [77], or [81] for complete details. The details are very similar to the proof of Theorem 3.4.2 below.

Exercises for Sect. 3.1

1. Let R and S be rings of characteristic zero. Let $\varphi : R \to S$ be an isomorphism. Show that φ restricts to an isomorphism $ML(R) \to ML(S)$. [First show that $\varphi^{-1} D \varphi \in$ LND A for every $D \in$ LND B.]

2. Let R and S be rings of characteristic zero. Let $\varphi : R \to S$ be an isomorphism. Show that φ restricts to an isomorphism $\mathscr{D}(R) \to \mathscr{D}(S)$.

3. Let k be a field. Let A be a k-algebra. Let $\{A_n\}$ be a $\mathbb{Z}$-filtration on A. Let $D \in$ Der A such that D respects the filtration. Let gr D be defined as in Definition 3.1.6.
   (i) Show that gr $D = 0$ if and only if $D = 0$.
   (ii) Show that if $D \in$ LND A, then gr $D \in$ LND(gr A).

4. (Maubach) Let $R = \mathbb{C}[X^2, X^3, Y^3, Y^4, Y^5, XY, X^2Y, XY^2, XY^3]$. This exercise will establish that $ML(R)$ is trivial while $\mathscr{D}(R)$ is not.
   (i) Show that $D_1 = Y^3 \partial_X$ and $D_2 = X^2 \partial_Y$ are locally nilpotent derivations on R, so that $ML(R)$ is trivial.
   (ii) Show that if $D \in$ LND R, then D extends uniquely to a locally nilpotent derivation on $\mathbb{C}[X, Y]$. [Use Propositions 1.2.15 and 1.3.37 from [117].]
   (iii) Let $f \in \mathbb{C}[X, Y]$ be a coordinate. Let $p(T) \in \mathbb{C}[T]$ such that $p(f) \in R$. Show that XY does not appear as a monomial in $p(f)$.
   (iv) Let $D \in$ LND R, and suppose there exists $g \in R^D$ such that XY appears as a monomial in g. Show that $D = 0$. [Use Rentschler's Theorem 5.1.12, [117].]
   (v) Show that $XY \notin \mathscr{D}(R)$, so that $\mathscr{D}(R)$ is not trivial.

3.2 Exponential Maps

If k is a field with prime characteristic p, we can certainly study derivations on a k-algebra. However, the kernel of any derivation will contain the pth power of every element, and so our earlier notion of "trivial" Makar-Limanov invariant becomes muddied by these pth powers. In this setting, the kernel of a derivation simply does not convey the appropriate information. As discussed in §9.5 [117], recall that for the coordinate ring of a variety over a field of characteristic zero, locally nilpotent derivations on the ring are equivalent to additive group actions on the variety. In this case, the kernel of a locally nilpotent derivation is equal to the ring of invariants of the corresponding additive group action. From another point of view, in characteristic zero, a locally nilpotent derivation D is equivalent to an exponential map $\exp TD = \sum \frac{1}{n!} T^n D^n$ (See §1.2, [117]). Using this perspective we can redefine the Makar-Limanov invariant for prime characteristic k-algebras. While we will focus on k-algebras, note that the following concepts can be defined for more general rings as well.

Definition 3.2.1 Let k be a field. Let A be a k-algebra and $\delta : A \to A^{[1]}$ a homomorphism of k-algebras. We write $\delta = \delta_T : A \to A[T]$ if we wish to emphasize an indeterminate T. We call δ an **exponential map** on A if

1. $\varepsilon_0 \delta_T$ is the identity on A, where $\varepsilon_0 : A[T] \to A$ is evaluation at $T = 0$, and
2. $\delta_S \delta_T = \delta_{S+T}$, where δ_S is extended to a homomorphism $A[T] \to A[S, T]$ by $\delta_S(T) = T$.

Let EXP A denote the set of all exponential maps on A. For $\delta \in$ EXP A, define $A^\delta = \{a \in A \mid \delta(a) = a\}$, a subalgebra of A called the **ring of δ-invariants**. For $a \in A$, write

$$\delta(a) = \sum_{n=0}^{\infty} \delta^{(n)}(a) T^n.$$

Since δ is a homomorphism, we see that $\{\delta^{(n)}\}_{n \geq 0}$ is a sequence of linear maps on A. This sequence is a **locally finite iterative higher derivation**, sometimes atrociously abbreviated as **lfihd**. It plays the prime characteristic role of a locally nilpotent derivation.

Definition 3.2.2 Let R be a ring. A **higher derivation** on R is a sequence $D = \{D^{(n)}\}_{n \geq 0}$ of linear maps $D^{(n)} : R \to R$, $n \geq 0$, such that

1. $D^{(0)}$ is the identity on R, and
2. (**Leibniz rule**) $D^{(n)}(rs) = \displaystyle\sum_{i+j=n} D^{(i)}(r) D^{(j)}(s)$ for all $n \geq 0$ and all $r, s \in R$.
   The higher derivation D is called **locally finite** if
3. $\{D^{(n)}(r)\}_{n \geq 0}$ consists of finitely many nonzero terms for each $r \in R$.
   The higher derivation D is called **iterative** if
4. $D^{(i)} D^{(j)} = \dbinom{i+j}{i} D^{(i+j)}$ for all $i, j \geq 0$.

Returning to our exponential map $\delta \in$ EXP A, since δ is a homomorphism of k-algebras, we see for each n that the map $\delta^{(n)} : A \to A$ is linear and that the Leibniz rule holds. Moreover, referring back to the definition of exponential map, property 1 in the definition is equivalent to saying $\delta^{(0)}$ is the identity on A, and property 2 is equivalent to the iterative property. Finally, since $\delta(a)$ is a polynomial in T, the sequence $\{\delta^{(n)}(a)\}_{n \geq 0}$ has finitely many nonzero elements for each $a \in A$, and thus the locally finite property is also satisfied. So $\{\delta^{(n)}\}_{n \geq 0}$ is indeed a locally finite iterative higher derivation on A. Conversely, a lfihd D on A determines an exponential map on A given by $\sum_{n \geq 0} T^n D^{(n)}$.

For $\delta \in \mathrm{EXP}\, A$, note that $\delta^{(1)} \in \mathrm{LND}\, A$. When the characteristic of A is zero, we can use the iterative property to write

$$\delta^{(n)} = \frac{1}{n!}\left(\delta^{(1)}\right)^n$$

for each $n \geq 0$. Then $A^\delta = \ker \delta^{(1)}$ and $\delta = \exp T \delta^{(1)}$. However, when the characteristic of A is not zero, then $\delta^{(1)}$ alone does not carry all the information about δ.

Definition 3.2.3 Let k be a field. Let A be a k-algebra. The **Makar-Limanov invariant** of A is

$$\mathrm{ML}(A) = \bigcap_{\delta \in \mathrm{EXP}\, A} A^\delta.$$

If X is an algebraic variety with coordinate ring $A(X)$, define $\mathrm{ML}(X) = \mathrm{ML}(A(X))$.

Note that the trivial exponential map is the standard inclusion $a \mapsto a$, and $\mathrm{ML}(A) = A$ if and only if the trivial map is the only exponential map on A. We call A **rigid** if $\mathrm{ML}(A) = A$.

Example 3.2.4 If k is a field, then $\mathrm{ML}(k^n) = \mathrm{ML}(k^{[n]}) = k$ for $n \geq 1$, because there are n "partial derivative" exponential maps δ_i, $i = 1, \ldots, n$, given by $\delta_i(x_j) = x_j + \delta_{ij} T$, where δ_{ij} is the Kronecker delta.

Definition 3.2.5 Let k be a field. Let A be a k-algebra. The **Derksen invariant** of A, denoted $\mathscr{D}(A)$, is the subalgebra of A generated by the rings of invariants of all nontrivial exponential maps on A. If X is an algebraic variety with coordinate ring $A(X)$, define $\mathscr{D}(X) = \mathscr{D}(A(X))$.

Example 3.2.6 If k is a field, then $\mathscr{D}(k^n) = \mathscr{D}(k^{[n]}) = k^{[n]}$ for $n \geq 2$.

As with locally nilpotent derivations in characteristic zero, we gain more useful tools when we restrict to integral domains. Sometimes we will need to require an integral domain to be finitely generated over a field k. Such an integral domain is called an **affine domain** and represents the ring of regular functions of an irreducible affine variety.

Definition 3.2.7 Given an integral domain A over a field k and $\delta \in \mathrm{EXP}\, A$, we define the δ-**degree** of $0 \neq a \in A$ by

$$\deg_\delta(a) = \deg_T(\delta(a)) = \max\{n \mid \delta^{(n)}(a) \neq 0\}$$

and we define $\deg_\delta(0) = -\infty$.

This is a degree function on A, and we can use it to prove prime characteristic analogues of several basic facts from §1.3, [117]. First note that A^δ is factorially closed in A by virtue of this degree function, and hence A^δ is also algebraically closed in A. Next observe that from the iterative property we have

$$\deg_\delta(\delta^{(n)}(a)) \leq \deg_\delta(a) - n$$

for each $a \in A$ and each $n \geq 0$. In particular, this shows that if $0 \neq a \in A$ and $\deg_\delta(a) = n$, then $\delta^{(n)}(a) \in A^\delta$. To keep going with the analogy, we need a notion of preslice. Assuming δ is nontrivial, we call $x \in A$ a **preslice** of δ if

$$\deg_\delta(x) = \min\{\deg_\delta(a) \mid a \in A \setminus A^\delta\}.$$

Let $x \in A$ be a preslice of δ and let $n = \deg_\delta(x)$. (Of course, if the characteristic of A is zero, then we have $n = 1$.) We call x a **slice** of δ if $\delta^{(n)}(x) = 1$.

Lemma 3.2.8 *Let A be an integral domain over a field k with characteristic $p \geq 0$. Let $\delta \in \mathrm{EXP}\, A$ be nontrivial. Let $x \in A$ be a preslice of δ. Then*

(i) $\delta^{(i)}(x) \in A^\delta$ *for all $i \geq 1$.*
(ii) $\delta^{(i)}(x) = 0$ *whenever $i > 1$ is not a power of p. In particular, $\deg_\delta(x)$ is a power of p.*
(iii) *If $0 \neq a \in A$, then $\deg_\delta(x)$ divides $\deg_\delta(a)$.*

Proof Let $n = \deg_\delta(x)$. Since $\deg_\delta(\delta^{(i)}(x)) \leq n - i$, for $i \geq 1$ we have $\deg_\delta(\delta^{(i)}(x)) \leq 0$ by the minimality of n. So $\delta^{(i)}(x) \in A^\delta$ for $i \geq 1$, proving (i). If $p = 0$, then the entire proof is done. Now suppose p is prime. Let $i > 1$, not a power of p. Write $i = p^j q$ where $j \geq 0$ and $q \geq 2$ is an integer not divisible by p. Note then $\delta^{(i-p^j)}(x) \in A^\delta$ by (i). Making use of the identity

$$\binom{p^j q}{p^j} \equiv q \pmod{p}$$

we have

$$q\delta^{(i)}(x) = \binom{i}{p^j}\delta^{(i)}(x) = \delta^{(p^j)}\delta^{(i-p^j)}(x) = 0.$$

Dividing by q we have $\delta^{(i)}(x) = 0$, and this completes the proof of (ii). Continuing with (iii) in the case p is prime, write $n = p^m$ for some integer $m \geq 0$. If $m = 0$ we are done, so assume now $m > 0$. Let $a \in A$ and let $r = \deg_\delta(a)$. Again, if $r = 0$ we are done, so assume $r > 0$. We must show $p^m \mid r$. First suppose $p \nmid r$. Then $\delta^{(1)}\delta^{(r-1)}(a) = r\delta^{(r)}(a) \neq 0$.

Now $\deg_\delta(\delta^{(r-1)}(a)) \le r - (r-1) = 1$, and so $\deg_\delta(\delta^{(r-1)}(a)) = 1 < n$. This contradicts the minimality of n. So we must have $p \mid r$. Now we can write $r = p^l r$ where $l \ge 1$ and $p \nmid r$. Again using the above binomial identity, we see $\binom{r}{p^l}$ is not divisible by p, and so

$$\delta^{(p^l)} \delta^{(r-p^l)}(a) = \binom{r}{p^l} \delta^{(r)}(a) \ne 0.$$

Since $\deg_\delta(\delta^{(r-p^l)}(a)) \le r - (r - p^l) = p^l$, this shows that $\deg_\delta(\delta^{(r-p^l)}(a)) = p^l$. By the minimality of $n = p^m$, we must have $l \ge m$, and so $n \mid r$. $\square$

Lemma 3.2.9 *Let A be an integral domain over a field k with characteristic $p \ge 0$. Let $\delta \in \text{EXP } A$ be nontrivial. Let $x \in A$ be a preslice of δ. Let $n = \deg_\delta(x)$ and let $d = \delta^{(n)}(x)$. Then $A \subseteq A^\delta[d^{-1}][x]$.*

Proof Let $0 \ne a \in A$. Let $r = \deg_\delta(a)$ and write $r = nm$ for some $m \in \mathbb{Z}$ by Lemma 3.2.8 (ii). We proceed by induction on m to show $a \in A^\delta[d^{-1}][x]$. If $m = 0$, then $a \in A^\delta \subseteq A^\delta[d^{-1}][x]$. Now suppose $m \ge 1$. Note that $\deg_\delta(x^m) = r$ and $\delta^{(r)}(x^m) = d^m$. Let $y = d^m a - \delta^{(r)}(a)x^m$. Now $\delta^{(r)}(y) = 0$ and so $\deg_\delta(y) < r$. Hence $\deg_\delta(y) \le n(m-1)$ by Lemma 3.2.8 (iii). So $y \in A^\delta[d^{-1}][x]$ by the induction hypothesis. Now $d^m a = y + \delta^{(r)}(a)x^m \in A^\delta[d^{-1}][x]$. Dividing by d^m we have $a \in A^\delta[d^{-1}][x]$ as desired. $\square$

Corollary 3.2.10 *Let A be an integral domain over a field k with characteristic $p \ge 0$. Let $\delta \in \text{EXP } A$. If δ admits a slice $s \in A$, then $A = A^\delta[s]$.*

Keeping in mind that A^δ is algebraically closed in A, we have

Corollary 3.2.11 *Let k be a field with characteristic $p \ge 0$. Let A be an integral domain over k such that $\text{trdeg}_k Q(A)$ is finite. Let $\delta \in \text{EXP } A$ be nontrivial. Then $\text{trdeg}_k Q(A^\delta) = \text{trdeg}_k Q(A) - 1$.*

Corollary 3.2.12 *Let k be a field with characteristic $p \ge 0$. Let A be an integral domain over k with $\text{trdeg}_k Q(A) = 1$. Let $\delta \in \text{EXP } A$ be nontrivial. Then $k \subseteq A^\delta$ is an algebraic field extension and δ admits a slice. In particular, if k is algebraically closed, then $A = k^{[1]}$.*

Lemma 3.2.13 *Let A be an affine domain over a field k with characteristic $p \ge 0$. Let $\delta \in \text{EXP } A$ be nontrivial. Let $x \in A$ be a preslice of δ. Let $n = \deg_\delta(x)$ and let $d = \delta^{(n)}(x)$. Then there exists $\tilde{\delta} \in \text{EXP } A$ with $A^{\tilde{\delta}} = A^\delta$ and $\deg_{\tilde{\delta}}(x) = 1$.*

Proof Let $\{a_i\}$ be a finite set of generators for A. Then $a_i \in A^\delta[d^{-1}][x]$ for each i by Lemma 3.2.9. Take $m \geq 0$ sufficiently large so that $d^m a_i \in A^\delta[x]$ for each i. Define $\tilde{\delta}(x) = x + d^m T$ and $\tilde{\delta}(a) = a$ for all $a \in A^\delta$. Extend $\tilde{\delta}$ to a homomorphism $\tilde{\delta} : A^\delta[d^{-1}][x] \to A^\delta[d^{-1}][x][T]$ to obtain an exponential map on $A^\delta[d^{-1}][x]$ with ring of invariants $A^\delta[d^{-1}]$. By the choice of m, we have $\tilde{\delta}(A) \subset A[T]$. So we can restrict $\tilde{\delta}$ to an exponential map on A with $A^{\tilde{\delta}} = A^\delta$. $\qquad\square$

We now return to filtrations and associated graded rings. Let A be an affine domain over a field k with characteristic $p \geq 0$. Suppose A admits a proper $\mathbb{Z}$-filtration $\{A_n\}_{n \in \mathbb{Z}}$ and let

$$\text{gr } A = \bigoplus_{n \in \mathbb{Z}} A_n/A_{n-1}$$

be the associated graded domain. Let Γ be a finite set of generators of A. We say the $\mathbb{Z}$-filtration is **admissible** under Γ if for each $n \in \mathbb{Z}$ it is possible to write any $a \in A_n$ as a finite sum of monomials in elements of Γ with each of these monomials belonging to A_n. The purpose of this assumption is to be able to obtain a finite set of homogeneous generators for $\text{gr } A$. Indeed, let $\rho : A \to \text{gr } A$ be the map defined by

$$\rho(a) = a + A_{n-1} \text{ for all } a \in A_n \setminus A_{n-1} \qquad (3.2.1)$$

sending each element $a \in A$ to its homogeneous **top part** in $\text{gr } A$. It is common to write $\text{gr } a$ instead of $\rho(a)$ when working with these homogeneous elements. If the filtration on A is admissible under Γ, then $\text{gr } A$ is generated by $\rho(\Gamma)$. Under this condition we will be able to homogenize an exponential map as we did with derivations in the previous section.

Suppose $\delta : A \to A[T]$ is a nontrivial exponential map on A. Let Γ be a finite set of generators of A. Let deg be the degree function determined by the filtration on A. Define

$$\deg(T) = \min \left\{ \frac{\deg(a) - \deg(\delta^{(i)}(a))}{i} \;\middle|\; a \in \Gamma, i \in \mathbb{Z}^+ \right\} \qquad (3.2.2)$$

If the filtration on A is admissible under Γ, then $\deg(T)$ is a rational number. Then we can extend deg to a degree function on $A[T]$ and we have $\deg(\delta^{(n)}(a)T^n) \leq \deg(a)$ for all $a \in A$ and all $n \geq 0$. For each $a \in A$ define

$$S(a) = \{n \mid \deg(\delta^{(n)}(a)T^n) = \deg(a)\}$$

and define

$$\text{gr}(\delta)(\rho(a)) = \sum_{n \in S(a)} \rho(\delta^{(n)}(a))T^n.$$

We can extend this linearly to obtain a map $\mathrm{gr}\,\delta : \mathrm{gr}\,A \to \mathrm{gr}(A)[T]$. In fact the map $\mathrm{gr}\,\delta$ is an exponential map on $\mathrm{gr}\,A$ which we call the **homogenization** of δ.

Theorem 3.2.14 (Derksen, Hadas, Makar-Limanov) *Let A be an affine domain over a field k with characteristic $p \geq 0$. Let $\delta \in \mathrm{EXP}\,A$ be nontrivial. Suppose there is a $\mathbb{Z}$-filtration on A which is admissible. Then the induced map $\mathrm{gr}\,\delta$ defined above is a nontrivial exponential map on $\mathrm{gr}\,A$ such that $\rho(A^{\delta}) \subseteq \mathrm{gr}(A)^{\mathrm{gr}\,\delta}$.*

Homogenization of an exponential map is an essential tool for computing the Derksen and Makar-Limanov invariants. We will use this tool in the following sections to prove a cancellation theorem for surfaces and to give cancellation counterexamples for threefolds. Before moving to the Cancellation Problem, we briefly mention a couple of other interesting results concerning exponential maps, one older and one more recent.

Recall Rentschler's Theorem (5.1.12, [117]) which states that for a field k with characteristic zero every locally nilpotent derivation of $k[X_1, X_2]$ is equivalent (up to conjugation by an automorphism) to a derivation of the form $f(X_2)\partial_{X_1}$. In [85] Masayoshi Miyanishi proved the prime characteristic version of Rentschler's result.

Theorem 3.2.15 (Miyanishi) *Let k be a field with prime characteristic p. Every exponential map of $k[X_1, X_2]$ is equivalent up to conjugation by an automorphism to a map δ such that*

$$\delta(X_1) = X_1 + f_0(X_2)T + f_1(X_2)T^p + f_2(X_2)T^{p^2} + \cdots + f_n(X_2)T^{p^n}$$

and $\delta(X_2) = X_2$ for some $n \geq 0$ and some $f_0, \ldots, f_n \in k[X_2]$.

A more recent result from Emilie Dufresne and Andreas Maurischat concerns Hilbert's 14th problem. Recall from §9.6, [117] that counterexamples to Hilbert's 14th problem in dimensions $n = 5, 6, 7$ have been constructed using locally nilpotent derivations on $k^{[n]}$ when k has characteristic zero. The examples are due to Daniel Daigle and Gene Freudenburg ($n = 5$, see [18]), Freudenburg ($n = 6$, see [47]), and Paul Roberts ($n = 7$, see [32, 104]). See Theorems 9.6.2 and 9.6.17 in [117] for the specific derivations in $k^{[7]}$ and $k^{[5]}$, respectively. By rescaling variables where necessary, in [39] Dufresne and Maurischat provided characteristic free formulations of these examples as follows. In each example, k is a field and δ is an exponential map.

Example 3.2.16 (Daigle, Freudenburg) Define an exponential map δ on $k[X_1, X_2, X_3, X_4, X_5])$ by

$$\delta(X_1) = X_1, \quad \delta(X_2) = X_2 + X_1^3 T$$
$$\delta(X_3) = X_3 + 2X_2 T + X_1^3 T^2, \quad \delta(X_4) = X_4 + 3X_3 T + 3X_2 T^2 + X_1^3 T^3$$
$$\delta(X_5) = X_5 + X_1^2 T$$

Example 3.2.17 (Freudenburg) Define an exponential map δ on $k[X_1, X_2, X_3, X_4, X_5, X_6]$ by

$$\delta(X_1) = X_1, \quad \delta(X_2) = X_2$$
$$\delta(X_3) = X_3 + X_1^3 T, \quad \delta(X_4) = X_4 + 2X_2^3 X_3 T + X_1^3 X_2^3 T^2$$
$$\delta(X_5) = X_5 + 3X_2^3 X_4 T + 3X_2^6 X_3 T^2 + X_1^3 X_2^6 T^3, \quad \delta(X_6) = X_6 + X_1^2 X_2^2 T$$

Example 3.2.18 (Roberts) Define an exponential map δ on $k[X_1, X_2, X_3, X_4, X_5, X_6, X_7]$ by

$$\delta(X_1) = X_1, \quad \delta(X_2) = X_2, \quad \delta(X_3) = X_3$$
$$\delta(X_4) = X_4 + X_1^3 T, \quad \delta(X_5) = X_5 + X_2^3 T, \delta(X_6) = X_6 + X_3^3 T$$
$$\delta(X_7) = X_7 + (X_1 X_2 X_3)^2 T$$

For each of these examples Dufresne and Maurischat proved that the ring of invariants is finitely generated when the characteristic of k is prime. Finite generation for Robert's example was also shown earlier by Kazuhiko Kurano in [66] using a different, less elementary approach. Remarkably, in contrast to the characteristic zero setting, it appears there are still no prime characteristic examples of exponential maps on $k^{[n]}$ with infinitely generated ring of invariants.

Exercises for Sect. 3.2

1. Let A be an integral domain over a field k. Let $\delta \in \mathrm{EXP}\,A$. Show that A^δ is both factorially closed in A and algebraically closed in A. [Use δ-degree.]
2. Let A be an integral domain over a field k of characteristic $p \geq 0$. Let $\delta \in \mathrm{EXP}\,A$ with corresponding locally finite iterative higher derivation $\{\delta^{(n)}\}_{n \in \mathbb{Z}}$. Let $a \in A$.
   (i) Show that $\deg_\delta(\delta^{(n)}(a)) \leq \deg_\delta(a) - n$ for all $n \geq 0$.
   (ii) Suppose $\delta^n(a) \neq 0$, $p \nmid (n+1)$, and $p \nmid \deg_\delta(\delta^n(a))$ for some $n \geq 0$. Show that $\deg_\delta(\delta^{(n)}(a)) = \deg_\delta(a) - n$.
   (iii) Suppose $\delta^n(a) \neq 0$, $p \nmid (n+1)$, and $p \mid \deg_\delta(\delta^n(a))$ for some $n \geq 0$. Show that $\deg_\delta(\delta^{(n)}(a)) < \deg_\delta(a) - n$.
   (iv) Give counterexamples to show (ii) and (iii) do not hold if $p \mid (n+1)$. [Consider $k^{[1]} = k[X]$ and define $\delta \in \mathrm{EXP}\,k[X]$ by $\delta(X) = X + T + T^p$.]
3. (Miyanishi) Let A be an integral domain over a field k of characteristic $p > 0$. Let $\delta \in \mathrm{EXP}\,A$ with corresponding locally finite iterative higher derivation $\{\delta^{(n)}\}_{n \in \mathbb{Z}}$. Let $n \geq 0$ and let $n = \sum_{i=0}^m n_i p^i$ be the p-adic expansion of n. Show that

$$\delta^{(n)} = \frac{1}{(n_0)!(n_1)! \cdots (n_m)!} \left(\delta^{(1)}\right)^{n_0} \left(\delta^{(p)}\right)^{n_1} \cdots \left(\delta^{(p^m)}\right)^{n_m}$$

4. (Hadas) Let A be an integral domain over a field k. Let $\delta \in \mathrm{EXP}\,A$, and let $\delta : Q(A) \to Q(A)[T]$ also denote the extension of δ to $Q(A)$.

(i) Let $a, b \in A$ with $b \neq 0$, and suppose $\frac{a}{b} \in Q(A)^{\delta}$. Let $n \geq 0$. Show that if
$\delta^{(n)}(b) \neq 0$, then $\frac{a}{b} = \frac{\delta^{(n)}(a)}{\delta^{(n)}(b)}$.
(ii) Show that $Q(A)^{\delta} = Q(A^{\delta})$.

5. (Makar-Limanov) Let A be an integral domain over a field k of characteristic $p \geq 0$.
Let $n, m \in \mathbb{Z}$ such that $n, m \geq 2$ and neither is a power of p. Let $\delta \in \operatorname{EXP} A$, and
let $c, d \in A^{\delta}$, both nonzero. Let $a, b \in A$. Show that if $0 \neq ca^n + db^m \in A^{\delta}$, then
$a, b \in A^{\delta}$.

6. Give a counterexample to show that Exercise 5 does not hold if either n or m is a power
of p.

7. Let A and B be algebras over a field k. Let $\varphi : A \to B$ be an isomorphism. Show that
φ restricts to an isomorphism $\operatorname{ML}(A) \to \operatorname{ML}(B)$. [First show that $\varphi^{-1}\delta\varphi \in \operatorname{EXP} A$ for
every $f \in \operatorname{EXP} B$.]

8. Let A and B be algebras over a field k. Let $\varphi : A \to B$ be an isomorphism. Show that
φ restricts to an isomorphism $\mathscr{D}(A) \to \mathscr{D}(B)$.

9. Let k be an algebraically closed field with characteristic $p \neq 2, 3$, or 5. Let $A =
k[X, Y, Z]/(X^2 + Y^3 + Z^5)$. Write $x, y, z \in A$ for the congruence classes of X, Y, Z,
respectively. This exercise will establish that A is rigid, that is, $\operatorname{ML}(A) = A$.
(i) Use the weights $w(x) = 15$, $w(y) = 10$, $w(z) = 6$ to define a $\mathbb{Z}$-filtration on A.
Show that the filtration is admissible and $\operatorname{gr} A \cong A$.
(ii) Suppose $\delta \in \operatorname{EXP} A$ is nontrivial. Let $\operatorname{gr} \delta$ be the homogenization of δ. Show that
$\operatorname{gr}(A)^{\operatorname{gr} \delta}$ contains a homogeneous element $f(y, z) \in k[y, z]$ with $f(y, z) \notin k$. [If
$g \in A^{\delta}$, write $g = xf_1(y, z) + f_2(y, z)$ for some $f_1(y, z), f_2(y, z) \in k[y, z]$. Show
that either $\operatorname{gr} g = \operatorname{gr}(xf_1(y, z))$ or $\operatorname{gr} g = \operatorname{gr}(f_2(y, z))$. In the first case, consider
$(\operatorname{gr} g)^2$.]
(iii) Suppose $\delta \in \operatorname{EXP} A$ is nontrivial. Let $\operatorname{gr} \delta$ be the homogenization of δ. Let
$f(y, z) \in \operatorname{gr}(A)^{\operatorname{gr} \delta}$ be homogeneous, with $f(y, z) \notin k$, as above. By factoring
$f(y, z)$ and considering cases, show that $\operatorname{gr}(A)^{\operatorname{gr} \delta}$ contains at least one of y, z, or
$\lambda y^3 + \mu z^5$ for some $\lambda, \mu \in k^*$. Conclude that $\operatorname{gr} \delta$ is trivial. [Use Exercises 1 and
5 above.]
(iv) Show that $\operatorname{ML}(A) = A$.

3.3   Cancellation in Dimensions One and Two

The Makar-Limanov and Derksen invariants have been used successfully to study
cancellation problems, such as the statement given in 9.4.10, [117]. In this section we
will reexamine this problem in a characteristic free setting. Let k denote an algebraically
closed field with characteristic $p \geq 0$. From an algebraic perspective, 9.4.10, [117] can be
restated as follows. (See Sect. 2.2 Exercise 3, [117].)

Cancellation Problem *Let k be an algebraically closed field. Let A be an affine domain over k. Suppose $A^{[1]} \cong k^{[n+1]}$ for some $n \geq 1$. Does it follow that $A \cong k^{[n]}$?*

Note that in the statement of the Cancellation Problem $A = A(V)$ for some irreducible affine variety V with $\dim V = n$. The dimension of V coincides with the transcendence degree of $Q(A)$ over k. When $n = 1$, we have $\mathrm{trdeg}_k Q(A) = 1$ and V is a curve. The affirmative solution for $n = 1$ is due to Shreeram Abhyankar, Paul Eakin, and William Heinzer. More generally, in [1] they proved

Theorem 3.3.1 (Abhyankar, Eakin, Heinzer) *Let k be an algebraically closed field. Let A and B be affine domains over k with $\mathrm{trdeg}_k Q(A) = \mathrm{trdeg}_k Q(B) = 1$. If $A^{[m]} \cong B^{[m]}$ for some positive integer m, then $A \cong B$.*

Anthony Crachiola and Leonid Makar-Limanov used the Makar-Limanov invariant to give a different proof of Theorem 3.3.1. In [16] they proved

Theorem 3.3.2 (Crachiola, Makar-Limanov) *Let k be an algebraically closed field. Let A be an affine domain over k with $\mathrm{trdeg}_k Q(A) = 1$, and suppose $A \not\cong k^{[1]}$. Let B be an integral domain over k. Then*

$$\mathrm{ML}(A \otimes_k B) = \mathrm{ML}(A) \otimes_k \mathrm{ML}(B).$$

Note that A in Theorem 3.3.2 is rigid, that is, $\mathrm{ML}(A) = A$, by Corollary 3.2.12. Taking $B = k^{[m]}$ we have

Corollary 3.3.3 *Let k be an algebraically closed field. Let A be an affine domain over k with $\mathrm{trdeg}_k Q(A) = 1$. Then $\mathrm{ML}(A^{[m]}) = \mathrm{ML}(A)$ for all $m \geq 1$.*

Using Theorem 3.3.2 we can quickly prove Theorem 3.3.1.

Proof of Theorem 3.3.1 First suppose neither A nor B is isomorphic to $k^{[1]}$. Using Corollary 3.2.12 and Theorem 3.3.2 we have

$$A = \mathrm{ML}(A) = \mathrm{ML}(A^{[m]}) \cong \mathrm{ML}(B^{[m]}) = \mathrm{ML}(B) = B.$$

Suppose instead that $A \cong k^{[1]}$. Then

$$\mathrm{ML}(B) = \mathrm{ML}(B^{[m]}) \cong \mathrm{ML}(A^{[m]}) = \mathrm{ML}(k^{[m+1]}) = k.$$

So B admits a nontrivial exponential map. Therefore $A \cong B \cong k^{[1]}$ by Corollary 3.2.12.

$\square$

We now turn to the Cancellation Problem for $n = 2$ which is a question about surfaces. As discussed in Sect. 9.4, [117], the work of Takao Fujita, Masayoshi Miyanishi, and Tohru Sugie (see [50, 88]) gives a positive solution to the Cancellation Problem for $n = 2$ when k has characteristic zero. Peter Russell extended their work to prime characteristic fields in [106]. More generally, combining their results we have

Theorem 3.3.4 (Fujita, Miyanishi, Russell, Sugie) *Let k be an algebraically closed field. Let A be an affine domain over k with $\mathrm{trdeg}_k Q(A) = 2$. If $A^{[m]} \cong k^{[m+2]}$ for some $m \geq 1$, then $A \cong k^{[2]}$.*

Their work uses some powerful ideas from algebraic geometry, including an important cancellation result due to Fujita and Shigeru Iitaka (see [62]) about varieties with logarithmic Kodaira dimension not equal to $-\infty$. In [17] Crachiola and Makar-Limanov showed how the Makar-Limanov invariant can be used to give a more self-contained, elementary proof of Theorem 3.3.4 in the special case $m = 1$ (i.e., the Cancellation Problem with $n = 2$). The first step is the following result about 2-dimensional rigid integral domains. The characteristic free proof below follows that given in [17]. Characteristic zero proofs can be found in [6, 78].

Theorem 3.3.5 (Crachiola, Makar-Limanov) *Let A be an affine domain over a field k. If $\mathrm{ML}(A) = A$, then $\mathrm{ML}(A^{[1]}) = A$.*

Proof Write $A^{[1]} = A[x]$ for an indeterminate x. Note that the A-linear map $A[x] \to A[x][T]$ defined by $x \mapsto x + T$ is an exponential map on $A[x]$ with ring of invariants A. Therefore $\mathrm{ML}(A[x]) \subseteq A$. To show $A \subseteq \mathrm{ML}(A[x])$ we must show that A is contained in the ring of invariants of every exponential map on $A[x]$.

Suppose $\delta : A[x] \to A[x][T]$ is an exponential map on $A[x]$ with $A \not\subseteq A[x]^\delta$. Now $A[x]$ is graded by x-degree, and the homogeneous elements are the monomials. In particular all nonzero elements of A are homogeneous. Write $\deg$ for the x-degree function. Let S be a finite generating set for A. Then $\Gamma = S \cup \{x\}$ is a finite generating set of homogeneous generators for $A[x]$ under which the filtration of $A[x]$ by x-degree is admissible. So we can define $\tau = \deg(T)$ as in Eq. (3.2.2). Note that there exists $s \in S$ with $s \notin A[x]^\delta$ by the assumption that $A \not\subseteq A[x]^\delta$. Now $\deg(s) = 0$ and so $\tau \leq 0$.

Let $\mathrm{gr}\,\delta$ be the nontrivial exponential map on $A[x]$ given by the homogenization of δ. Let us show that $x \in A[x]^{\mathrm{gr}\,\delta}$. Assuming not, then $\mathrm{gr}(\delta)^{(j)}(x) \neq 0$ for some $j \geq 1$. By the iterative property we have

$$\deg_{\mathrm{gr}\,\delta}(\mathrm{gr}(\delta)^{(j)}(x)) \leq \deg_{\mathrm{gr}\,\delta}(x) - j.$$

Since x is homogeneous we have $\mathrm{gr}(\delta)^{(j)}(x) = bx^m$, where $0 \neq b \in A$ and $m \geq 0$, and so

$$\deg_{\mathrm{gr}\,\delta}(b) + m \deg_{\mathrm{gr}\,\delta}(x) \leq \deg_{\mathrm{gr}\,\delta}(x) - j.$$

Thus

$$\deg_{\text{gr}\,\delta}(b) + (m-1)\deg_{\text{gr}\,\delta}(x) \le -j.$$

This implies $m = 0$, so $\text{gr}(\delta)^{(j)}(x) = b$. By construction of $\text{gr}\,\delta$, we have $1 = \deg(x) = \deg(bT^j) = j\tau$. But $\tau \le 0$, a contradiction. Therefore $x \in A[x]^{\text{gr}\,\delta}$.

Let $a \in A$ with $a \notin A[x]^{\text{gr}\,\delta}$. For $i \ge 0$, since a is homogeneous we have $\text{gr}(\delta)^{(i)}(a) = a_i x^{n_i}$, where $a_i \in A$ and $n_i \ge 0$. Then for each $a_i \ne 0$ we have $0 = \deg(a) = \deg(a_i x^{n_i} T^i) = n_i + i\tau$. Thus $n_i = -i\tau$ whenever $a_i \ne 0$. So $\text{gr}(\delta)(a) = \sum_i a_i(x^{-\tau}T)^i$. Therefore $\text{gr}\,\delta$ restricts to a nontrivial exponential map $A \to A[U]$, where $U = x^{-\tau}T$. This contradicts the hypothesis that $\text{ML}(A) = A$. $\square$

We can now use Theorem 3.3.5 to answer the Cancellation Problem for $n = 2$.

Theorem 3.3.6 *Let k be an algebraically closed field. Let A be an affine domain over k with $\text{trdeg}_k Q(A) = 2$. If $A^{[1]} \cong k^{[3]}$, then $A \cong k^{[2]}$.*

Proof As in the beginning of the proof of Theorem 3.3.5, we note that A is the ring of invariants of an exponential map on $A^{[1]}$. Thus A is factorially closed in $A^{[1]}$. Since $A^{[1]} \cong k^{[3]}$, we can view A as a factorially closed subalgebra of $k^{[3]}$. Thus A is a unique factorization domain with $\text{trdeg}_k Q(A) = 2$. Now $\text{ML}(A^{[1]}) = k$, and so there exists a nontrivial $\delta \in \text{EXP}\,A$ by Theorem 3.3.5. In the same manner, A^δ is a unique factorization domain, and since $\text{trdeg}_k Q(A^\delta) = 1$ we must have $A^\delta = k^{[1]}$. (See Sect. 3.3 Exercise 1.) Write $A^\delta = k[x]$ for some $x \in A^\delta$.

By Lemma 3.2.13 we can assume there exists $y \in A$ with $\delta(y) = y + f(x)T$, and where $f(x) \in A^\delta = k[x]$ has x-degree $\deg_x(f(x))$ as small as possible. Note that $k[x, y] \subseteq A$. We will show that $A = k[x, y]$. Suppose for the sake of contradiction there exists $a \in A$ with $a \notin k[x, y]$. By Lemma 3.2.9 we have $A \subset k(x)[y]$, so we can write a as a rational function with a numerator in $k[x, y]$ and denominator a nonconstant element of $k[x]$. Since k is algebraically closed, we can multiply a by some polynomial in $k[x]$ to obtain an element in A of the form $g(x, y) + h(y)/(x - \lambda)$ for some $g(x, y) \in k[x, y]$, $h(y) \in k[y]$, and $\lambda \in k$. So $h(y)/(x - \lambda)$ is an element of A. Now $h(y)$ splits into a product of linear factors in $k[y]$, and $x - \lambda$ is irreducible in A. Since A is a unique factorization domain we must have $(y - \mu)/(x - \lambda) \in A$ for some $\mu \in k$. Then

$$\delta\left(\frac{y-\mu}{x-\lambda}\right) = \frac{y-\mu}{x-\lambda} + \frac{f(x)}{x-\lambda}T.$$

But then $f(x)/(x - \lambda) \in A^\delta$ and $\deg_x(f(x)/(x - \lambda)) < \deg_x(f(x))$, contradicting our choice of $f(x)$. $\square$

Let us briefly mention a couple other 2-dimensional cancellation results before moving to higher dimensions. In an unpublished paper [19], Wlodzimierz Danielewski gave the following example for which cancellation fails. (See Sect. 9.4 Exercise 4, [117].)

Theorem 3.3.7 (Danielewski) *Let V and W be the surfaces in $\mathbb{C}^3$ given by $V : xy = z^2 - 1$ and $W : x^2y = z^2 - 1$. Then $V \times \mathbb{C} \simeq W \times \mathbb{C}$ but $V \not\simeq W$.*

Note that the surfaces given by Danielewski are not factorial. That is, the coordinate rings $A(V)$ and $A(W)$ are not unique factorization domains. Theorems 3.3.1 and 3.3.5 were used in [14] to prove

Theorem 3.3.8 (Crachiola) *Let k be an algebraically closed field. Let A and B be affine unique factorization domains over k with $\mathrm{trdeg}_k Q(A) = \mathrm{trdeg}_k Q(B) = 2$. If $A^{[1]} \cong B^{[1]}$, then $A \cong B$.*

Turning now to cancellation in higher dimensions, David Finston and Stefan Maubach found the following class of threefolds for which cancellation fails (see [45]).

Theorem 3.3.9 (Finston, Maubach) *Let a, b, c, m, n be positive integers with a, b, c pairwise relatively prime. Let $V_{n,m}$ be the threefold in $\mathbb{C}^5$ given by*

$$x^a + y^b + z^c = 0, \quad x^m u - y^n v = 1.$$

Then $V_{n,m} \times \mathbb{C} \simeq V_{n',m'} \times \mathbb{C}$ for all (m, n), (m', n'), but $V_{n,m} \not\simeq V_{n',m'}$ when $(m, n) \neq (m', n')$.

Finston and Maubach used a Danielewski-style argument to prove the isomorphism of cylinders over threefolds in the class, and they used the Makar-Limanov invariant to distinguish between members of the class. It is interesting to note that their threefolds are indeed factorial, although they are also singular.

More recently, Neena Gupta has made an impressive leap forward in the prime characteristic case (see [54–56]). Gupta applied the Derksen invariant to a result of Teruo Asanuma (see [5]) to show that the Cancellation Problem for $n \geq 3$ has a negative answer over a prime characteristic field. We examine her work in the following section.

Exercises for Sect. 3.3

1. Let k be an algebraically closed field. Let A be a unique factorization domain over k with $\mathrm{trdeg}_k Q(A) = 1$. Show that $A = k^{[1]}$. This result is used in the proof of Theorem 3.3.6. [Let $x \in A$ be irreducible. Show that if $y \in A$ is irreducible, then $y = \lambda x + \mu$ for some $\lambda, \mu \in k$ with $\lambda \neq 0$.]

2. Let k be an algebraically closed field. Let A be a unique factorization domain over k with $\mathrm{trdeg}_k Q(A) = 2$. Suppose $\mathrm{ML}(A) = k$. Show that $A = k^{[2]}$. [Use ideas from the proof of Theorem 3.3.6.]

3. Let k be an algebraically closed field. Let $A \neq k^{[1]}$ be an algebra over k with $\mathrm{trdeg}_k Q(A) = 1$. Let $n \geq 2$ and let $B = A^{[n]}$. Show that $\mathscr{D}(B)$ is trivial but $\mathrm{ML}(B)$ is not. [Use Corollary 3.3.3.]

3.4 Cancellation in Dimensions Three and Higher

As discussed in the previous section, the Cancellation Problem was settled for curves ($n = 1$) by Abhyankar, Eakin, and Heinzer (see [1]) in 1972 and for surfaces ($n = 2$) by the papers of Fujita, Miyanishi, Sugie, and Russell (see [50,88,106]) in 1980–1981. For $n \geq 3$ the problem remains open over fields with characteristic zero. Remarkably, three decades after the problem was settled for surfaces, the case $n \geq 3$ with prime characteristic has recently been answered negatively by Neena Gupta. In a 2014 paper [54] Gupta gave the first counterexample to the Cancellation Problem for $n = 3$. This work was then extended in two additional papers [55,56] published the same year to give a class of counterexamples for all $n \geq 3$.

Gupta's work built on an example of Asanuma that traces back to his 1987 paper [4]. While the work in that paper was more general, Asanuma revisited the subject in a 1994 paper [5] which focuses on the specific hypersurface V in k^4 given by

$$x^m y + z^{p^e} + t + t^{sp} = 0, \tag{3.4.1}$$

where $p > 0$ is the characteristic of k, and m, e, s are positive integers such that $p^e \nmid sp$ and $sp \nmid p^e$. Of note here is the following result of Asanuma. (See Theorem 5.1 in [4] and Theorem 1.1 in [5].)

Theorem 3.4.1 (Asanuma) *Let p be prime and let k be a field with characteristic p. Let m, e, s be positive integers such that $p^e \nmid sp$ and $sp \nmid p^e$. Let*

$$A = k[X, Y, Z, T]/(X^m Y + Z^{p^e} + T + T^{sp}).$$

Then $A^{[1]} \cong k^{[4]}$.

Asanuma found this result in the context of studying $\mathbb{A}^2$-fibrations. Briefly, let R be a Noetherian domain and let B be a finitely generated flat R-algebra. Let R_P denote the localization at a prime ideal P of R. Then B is called an $\mathbb{A}^n$-**fibration over** R if $B \otimes_R (R_P/PR_P) \cong (R_P/PR_P)^{[n]}$ for every prime ideal P of R. (Asanuma called this a **pseudopolynomial R-algebra in n variables**.) In other words the fibers of B over R are polynomial rings. The affine fibration problem of Igor Dolgachev and Boris

Weisfeiler (Veĭsfeĭler) asks if every $\mathbb{A}^n$-fibration B over R is necessarily isomorphic to $R^{[n]}$ (see [131]). While this is a deep problem with rich history, here we simply mention the connection to Asanuma's example. Avinash Sathaye proved in [108] that every $\mathbb{A}^2$-fibration over a principle ideal domain R containing $\mathbb{Q}$ must be isomorphic to $R^{[2]}$. The example of Asanuma demonstrated that the assumption $\mathbb{Q} \subset S$ could not be removed. Indeed, Asanuma showed that his example A is an $\mathbb{A}^2$-fibration over $k[x]$, but $A \neq k[x]^{[2]}$.

As Asanuma showed in [5], the corresponding hypersurface V in k^4 (with k infinite) admits an algebraic k^*-action given by

$$\lambda \cdot (x, y, z, t) = (\lambda x, \lambda^{-m} y, z, t)$$

which is not linearizable. This led Asanuma to an intriguing observation. If $A \cong k^{[3]}$, then he would have a prime characteristic counterexample to the linearization problem for k^3, while if $A \ncong k^{[3]}$, then he would have a prime characteristic counterexample to the Cancellation Problem for $n = 3$. This observation was dubbed "Asanuma's dilemma" by Russell (see Russell's problem 2 in [107]). In [54] Gupta settled Asanuma's dilemma as follows.

Theorem 3.4.2 (Gupta) *Let k and A be as above in Theorem 3.4.1. Also suppose $m > 1$ and suppose the field k is infinite. Let $x, y, z, t \in A$ denote the congruence classes of X, Y, Z, T, respectively. Then $y \notin \mathscr{D}(A)$. Therefore $A \ncong k^{[3]}$.*

The proof of Theorem 3.4.2 follows the same arguments used to show that the Koras–Russell hypersurface $x + x^2 y + z^2 + t^3 = 0$ in k^4 is not isomorphic to k^3. Along the way we will introduce multiple weights on x, y, z, t and each time consider the induced exponential map on the corresponding graded algebra. Before proceeding with the proof, we collect one more lemma which appears as Lemma 3.3 in [54].

Lemma 3.4.3 *Let k be an infinite field and let B be an affine domain over k. Let $f \in B$ such that $f - \gamma$ is prime in B for infinitely many $\gamma \in k$. Let $\varphi \in \mathrm{EXP}\, B$ with $f \in B^\varphi$. Then there exists $\beta \in k^*$ such that $f - \beta$ is prime in B and φ induces a nontrivial exponential map on $B/(f - \beta)$.*

Proof Let $b \in B$ with $b \notin B^\varphi$. Let $n = \deg_\varphi(b) \geq 1$. By the hypotheses there exists $\beta \in k^*$ such that $f - \beta$ is prime in B and $(f - \beta) \nmid \varphi^{(n)}(b)$ in B. Now $f - \beta \in B^\varphi$, so φ induces an exponential map $\overline{\varphi}$ on $B/(f - \beta)$. Let $\overline{b} \in B/(f - \beta)$ denote the image of b. Then $\overline{\varphi}^{(n)}(\overline{b}) \neq 0$ and so $\overline{\varphi}$ is nontrivial. $\square$

Proof of Theorem 3.4.2 Let $\delta \in \mathrm{EXP}\, A$ be nontrivial. We will show that $A^\delta \subseteq k[x, z, t]$. This means $y \notin A^\delta$ which implies $y \notin \mathscr{D}(A)$. We will assume that k is algebraically closed. For if k is not algebraically closed, then δ extends to an exponential map on the extension of scalars $A \otimes_k \bar{k}$, and if y is not invariant under this extension, then y is also not invariant under δ.

Write $s = qp^r$ where $q \nmid p$. Note that $q > 1$ since $sp \nmid p^e$. Also $r + 1 < e$ since $p^e \nmid sp$. View A as a subalgebra of $k[x, x^{-1}, z, t]$ by the relation $y = -x^{-m}(z^{p^e} + t + t^{sp})$. Introduce the following non-negative integer weights: $\deg_1(x) = 0$, $\deg_1(z) = q$, and $\deg_1(t) = p^{e-r-1}$. Note then $\deg_1(z^{p^e}) = qp^e$ and $\deg(t^{sp}) = sp \cdot p^{e-r-1} = qp^e$. Also note $\deg_1(t) < qp^e$. These weights make $k[x, x^{-1}, z, t]$ a $\mathbb{Z}$-graded algebra and induce an admissible $\mathbb{Z}$-filtration on A with associated graded ring

$$B = \operatorname{gr} A \cong k[X, Y, Z, T]/(X^m Y + Z^{p^e} + T^{sp}).$$

We will continue to write x, y, z, t for the congruence classes of X, Y, Z, T in B. By Theorem 3.2.14 δ induces a nontrivial exponential map $\operatorname{gr} \delta$ on B with $\operatorname{gr} f \in B^{\operatorname{gr} \delta}$ for every $f \in A^\delta$. Note that we can write $\deg_1(y) = qp^e$.

Now introduce weights $\deg_2(x) = -1$, $\deg_2(y) = m$, and $\deg_2(z) = \deg_2(t) = 0$. Under these weights we can view B as a $\mathbb{Z}$-graded algebra. We again apply Theorem 3.2.14, but let us continue to write $\operatorname{gr} \delta$ for the induced homogenized nontrivial exponential map on B. Note that for any $h \in B$ we can use the relation $x^m y = -(z^{p^e} + t^{sp})$ to write h uniquely as

$$h = h_0(x, z, t) + \sum_{\substack{j > 0 \\ 0 \leq i < m}} x^i y^j h_{ij}(z, t)$$

with $h_0(x, z, t) \in k[x, z, t]$ and $h_{ij}(z, t) \in k[z, t]$. Note that no two terms in this summation have the same degree under $\deg_2$. So under these new weights we have $\operatorname{gr} h = x^i y^j h_1(z, t)$ where $i, j \geq 0$ and $h_1(z, t) \in k[z, t]$.

Suppose, for the sake of a contradiction, that $A^\delta \not\subseteq k[x, z, t]$. Then there must be an element $f \in A^\delta$ such that $\operatorname{gr} f = x^a y^b f_1(z, t)$ for some integers a, b with $0 \leq a < m$, $b > 0$ and some $f_1(z, t) \in k[z, t]$. Since $B^{\operatorname{gr} \delta}$ is factorially closed in B, we must have $y \in B^{\operatorname{gr} \delta}$. By Corollary 3.2.11 we know $\operatorname{trdeg}_k Q(B^{\operatorname{gr} \delta}) = 2$, so there must exist some element $g \in A^\delta$ such that $\operatorname{gr} g \notin k[y]$. Write $\operatorname{gr} g = x^c y^d g_1(z, t)$ for some $c, d \geq 0$ and some $g_1(z, t) \in k[z, t]$. Since $B^{\operatorname{gr} \delta}$ is factorially closed in B, we have $x^c, g_1(z, t) \in B^{\operatorname{gr} \delta}$. Now if $g_1(z, t)$ is constant, then $\operatorname{gr} g \notin k[y]$ implies $c > 0$ and then $x \in B^{\operatorname{gr} \delta}$. In that case $z^{p^e} + t^{sp} = -x^m y \in B^{\operatorname{gr} \delta}$. So in either case, taking either $g_1(z, t)$ or $z^{p^e} + t^{sp}$, there exists some nonconstant polynomial $h(z, t) \in B^{\operatorname{gr} \delta}$ such that $k[y, h(z, t)] \subseteq B^{\operatorname{gr} \delta}$.

Now we return to the first set of weights: $\deg_1(x) = 0$, $\deg_1(y) = qp^e$, $\deg_1(z) = q$, and $\deg_1(t) = p^{e-r-1}$. We can assume $h(z, t)$ is homogeneous under these weights so that, since k is algebraically closed, $h(z, t)$ factors as

$$h(z, t) = \lambda z^i t^j \prod_l (z^{p^{e-r-1}} + \mu_l t^q)$$

for some $i, j \geq 0$ and some $\lambda, \mu_l \in k^*$. Since $h(z, t)$ is not constant and since $B^{\mathrm{gr}\,\delta}$ is factorially closed in B, there exists a prime element $w \in B^{\mathrm{gr}\,\delta}$ with either $w = z$, or $w = t$, or $w = z^{p^{e-r-1}} + \mu t^q$ for some $\mu \in k^*$. We will split these options into two cases. Each case will end in a contradiction.

Case one. Suppose $w = z^{p^{e-r-1}} + t^q$. Then $x^m y = -w^{p^{r+1}} \in B^{\mathrm{gr}\,\delta}$ which implies $x \in B^{\mathrm{gr}\,\delta}$. So we have $k[x, y, w] \subseteq B^{\mathrm{gr}\,\delta}$. Extend the scalar field to $L = k(x, y, w)$ to obtain the ring

$$D = B \otimes_{k[x,y,w]} L \cong L[z, t]/(z^{p^{e-r-1}} + t^q - w)$$

on which $\mathrm{gr}\,\delta$ induces a nontrivial exponential map, say φ. We will now show this is impossible. Let $\overline{L}$ denote an algebraic closure of L. Note that $t^q - w \in L[t]$ has no repeated roots in $\overline{L}$ since $p \nmid q$. So by Eisenstein's criterion $z^{p^{e-r-1}} + (t^q - w)$ is irreducible as a polynomial in z over $\overline{L}[t]$. Therefore

$$D \otimes_L \overline{L} \cong \overline{L}[z, t]/(z^{p^{e-r-1}} + t^q - w)$$

is an integral domain. From this we see D is an integral domain, and L must be algebraically closed in D.

We will now show that $D \otimes_L \overline{L}$ is not integrally closed. Let $\xi \in \overline{L}$ be a root of $z^{p^{e-r-1}} - w \in \overline{L}[z]$. Then $z^{p^{e-r-1}} - w = (z - \xi)^{p^{e-r-1}}$. From the relation $z^{p^{e-r-1}} + t^q - w = 0$ we then have

$$(z - \xi)^{p^{e-r-1}} + t^q = 0.$$

Recall $p \nmid q$, so $p^{e-r-1} \neq q$. If $p^{e-r-1} < q$, then we have

$$\left(\frac{z - \xi}{t}\right)^{p^{e-r-1}} - t^{q - p^{e-r-1}} = 0$$

which shows that $\frac{z-\xi}{t} \in Q(D \otimes_L \overline{L})$ is integral over $D \otimes_L \overline{L}$. Similarly, if $p^{e-r-1} > q$, then we find $\frac{t}{z-\xi} \in Q(D \otimes_L \overline{L})$ is integral over $D \otimes_L \overline{L}$. So $D \otimes_L \overline{L}$ is not integrally closed.

Returning now to the nontrivial exponential map φ on D, we have $D = (D^\varphi)^{[1]}$ by Corollary 3.2.12. Since D^φ is an algebraic extension of L and L is algebraically closed in D, we must have $D^\varphi = L$ and $D = L^{[1]}$. Now the nontrivial exponential map φ on D extends to a nontrivial exponential map on $D \otimes_L \overline{L}$ which implies $D \otimes_L \overline{L} \cong \overline{L}^{[1]}$. But $\overline{L}^{[1]}$ is integrally closed, bringing us to a contradiction. Therefore $w \neq z^{p^{e-r-1}} + t^q$, completing case one.

Case two. Suppose either $w = z$, or $w = t$, or $w = z^{p^{e-r-1}} + \mu t^q$ for some $\mu \in k^*$ with $\mu \neq 1$. Note that $w - \gamma$ is prime in $k[z, t]$ for all $\gamma \in k^*$ and hence $k[z, t]/(w - \gamma)$ is an integral domain for all $\gamma \in k^*$. Also, for $\gamma \in k^*$, we have $(w - \gamma) \nmid (z^{p^e} + t^{sp})$ in $k[z, t]$, and so $B/(w - \gamma)$ is an integral domain. By Lemma 3.4.3 there exists $\beta \in k^*$ such that gr δ induces a nontrivial exponential map on $B/(w - \beta)$. Note also that the ring of invariants of this exponential map contains the image of y in $B/(w - \beta)$ (which without confusion we continue to call y). Thus the nontrivial exponential map on $B/(w - \beta)$ extends to a nontrivial exponential map on

$$B/(w - \beta) \otimes_{k[y]} k(y) \cong k(y)[x, z, t]/(x^m y + z^{p^e} + t^{sp}, w - \beta).$$

Let $E = B/(w - \beta) \otimes_{k[y]} k(y)$. Similar to the argument in case one, we will show that E is not integrally closed. This will imply E cannot be a univariate polynomial ring over a field, contradicting the existence of a nontrivial exponential map on E. Let $v = z^{p^{e-r-1}} + t^q \in k[z, t]$ so that $v^{p^{r+1}} = z^{p^e} + t^{sp}$. So we have the relation

$$x^m y + v^{p^{r+1}} = 0$$

in E. Note that $v \neq 0$ and v is not a unit in E. If $p^{r+1} \leq m$, then we have

$$\left(\frac{v}{x}\right)^{p^{r+1}} + x^{m - p^{r+1}} y = 0$$

which shows that $v/x \in Q(E)$ is integral over E. Similarly, if $p^{r+1} > m$, then we find $x/v \in Q(E)$ is integral over E. So E is not integrally closed.

Having exhausted all possibilities, we have reached a contradiction to the assumption that $A^\delta \not\subseteq k[x, z, t]$ as desired. $\qquad\square$

Corollary 3.4.4 *Let k and A be as above in Theorem 3.4.1. Also suppose $m > 1$ and suppose the field k is infinite. Let $x, y, z, t \in A$ denote the congruence classes of X, Y, Z, T, respectively. Then*

(i) $\mathscr{D}(A) = k[x, z, t]$, *and*
(ii) $\mathrm{ML}(A) = k[x]$.

Proof First note we have exponential maps $\delta_1, \delta_2 : A \to A[U]$ defined as follows:

$$\delta_1(x) = x, \quad \delta_1(y) = y + U + x^{-m}(t^{sp} - (t - x^m U)^{sp})$$
$$\delta_1(z) = z, \quad \delta_1(t) = t - x^m U$$

and

$$\delta_2(x) = x, \quad \delta_2(y) = y + x^{m(p^e - 1)} U^{p^e}, \quad \delta_2(z) = z - x^m U, \quad \delta_2(t) = t.$$

Note that $A^{\delta_1} = k[x, z]$ and $A^{\delta_2} = k[x, t]$.

(i) From the proof of Theorem 3.4.2 we know $\mathscr{D}(A) \subseteq k[x, z, t]$. From the maps δ_1 and δ_2 we see that $k[x, z, t] \subseteq \mathscr{D}(A)$.

(ii) Since $A^{\delta_1} \cap A^{\delta_2} = k[x]$, we must have $\mathrm{ML}(A) \subseteq k[x]$. Let $\delta \in \mathrm{EXP}\, A$ be nontrivial. To obtain the reverse containment, we will show that $x \in A^{\delta}$. Suppose $x \notin A^{\delta}$. Let $f \in A^{\delta}$. Then $f \in k[x, z, t]$ by the proof of Theorem 3.4.2. Write $f = x f_1(x, z, t) + f_2(z, t)$ for some $f_1(x, z, t) \in k[x, z, t]$ and some $f_2(x, t) \in k[z, t]$. Since A^{δ} is factorially closed in A and since by assumption $x \notin A^{\delta}$, we must have $f_2(x, t) \neq 0$. Apply the weights $\deg_1$ and $\deg_2$ from the proof of Theorem 3.4.2 to obtain a nontrivial exponential map $\mathrm{gr}\,\delta$ on

$$B = \mathrm{gr}\, A \cong k[X, Y, Z, T]/(X^m Y + Z^{p^e} + T^{sp}).$$

Note that $\deg_2(x f_1(x, z, t)) < 0$, while $\deg_2(f_2(z, t)) = 0$. So with respect to $\deg_2$ we must have $\mathrm{gr}\, f = f_2(z, t) \in B^{\mathrm{gr}\,\delta}$. By Corollary 3.2.11 we know $\mathrm{trdeg}_k Q(A^{\delta}) = 2$, so let $g \in A^{\delta}$ be algebraically independent with f over k. Write $g = x g_1(x, z, t) + g_2(z, t)$ for some $g_1(x, z, t) \in k[x, z, t]$ and some $g_2(x, t) \in k[z, t]$. Again we conclude $0 \neq g_2(z, t) = \mathrm{gr}\, g \in B^{\mathrm{gr}\,\delta}$. Suppose f_2 and g_2 are algebraically dependent over k, say $P(f_2, g_2) = 0$ for some $P \in k^{[2]}$. Then $0 \neq P(f, g) \in A^{\delta}$, but x divides $P(f, g)$ in A. This contradicts the assumption that $x \notin A^{\delta}$. Hence $B^{\mathrm{gr}\,\delta}$ contains two algebraically independent elements of $k[z, t]$. Since $B^{\mathrm{gr}\,\delta}$ is algebraically closed in B, we conclude $k[z, t] \subseteq B^{\mathrm{gr}\,\delta}$. Now $x^m y = -(z^{p^e} + t^{sp}) \in B^{\mathrm{gr}\,\delta}$ which implies $x, y \in B^{\mathrm{gr}\,\delta}$. But then $\mathrm{gr}\,\delta$ is trivial which contradicts our hypothesis that δ is nontrivial. Therefore $x \in A^{\delta}$ and $\mathrm{ML}(A) = k[x]$.   □

In a follow-up paper to [54], Gupta considers the following more general class of threefolds given by

$$x^m y - F(x, z, t) = 0$$

in k^4, where $F \in k^{[3]}$. In [55] Gupta proves (among other things) the following two theorems.

Theorem 3.4.5 (Gupta)   *Let k be a field and let $F(X, Z, T) \in k[X, Z, T]$. Let*

$$A = k[X, Y, Z, T]/(X^m Y - F(X, Z, T))$$

with $m > 1$. Let $f(Z, T) = F(0, Z, T)$ and $G = X^m Y - F(X, Z, T)$. The following are equivalent:

1. *$f(Z, T)$ is a variable in $k[Z, T]$.*
2. *$A \cong_{k[x]} k[x]^{[2]}$, where x denotes the congruence class of X in A.*
3. *$A \cong k^{[3]}$.*

4. G is a variable in $k[X, Y, Z, T]$.
5. G is a variable in $k[X, Y, Z, T]$ along with X.

Theorem 3.4.5 includes not only Asanuma's example (3.4.1) but also the Koras–Russell hypersurface $x + x^2 y + z^2 + t^3 = 0$ in k^4. The equivalence of (1) and (3) in Theorem 3.4.5 precisely explains why neither can be isomorphic to k^3. The next result however does not include the Koras–Russell hypersurface.

Theorem 3.4.6 (Gupta) *Let k be a field and let $F(X, Z, T) \in k[X, Z, T]$. Let*

$$A = k[X, Y, Z, T]/(X^m Y - F(X, Z, T))$$

with $m \geq 1$. Let $f(Z, T) = F(0, Z, T)$. If $k[Z, T]/(f) \cong k^{[1]}$, then $A^{[1]} \cong_{k[x]} k[x]^{[3]} \cong k^{[4]}$.

Asanuma's threefold $x^m y + z^{p^e} + t + t^{sp} = 0$ in k^4 (with prime characteristic p) belongs to this class of threefolds with $F = Z^{p^e} + T + T^{sp}$. Indeed $k[Z, T]/(F) \cong k^{[1]}$. (See Sect. 5.3 Exercise 2 and Sect. 5.4 Exercise 5, [117].) However, F is not a variable in $k[Z, T]$, that is, $k[Z, T] \neq k[F]^{[1]}$ as was shown by Masayoshi Nagata (see page 39 in [94]). For this reason $F \in k[Z, T]$ is called a **nontrivial line**. In other words, F provides a counterexample to the prime characteristic version of the Epimorphism Theorem of Abhyankar, Moh, and Suzuki (Theorem 5.3.5, [117]).

Definition 3.4.7 Let k be a field and let $F(X, Z, T) \in k[X, Z, T]$. Let

$$A = k[X, Y, Z, T]/(X^m Y - F(X, Z, T))$$

with $m \geq 1$. Let $f(Z, T) = F(0, Z, T)$. We call A an **Asanuma threefold** if f is a nontrivial line in $k[Z, T]$.

As a consequence of these theorems, in [55] Gupta shows

Corollary 3.4.8 (Gupta) *Let p be prime and let k be a field with characteristic p. Let $F(X, Z, T) \in k[X, Z, T]$ and let $m > 1$. Let*

$$A = k[X, Y, Z, T]/(X^m Y - F(X, Z, T))$$

be an Asanuma threefold. Then $A^{[1]} \cong k^{[4]}$ while $A \ncong k^{[3]}$. That is, Asanuma threefolds with $m > 1$ are counterexamples to the Cancellation Problem with $n = 3$. Furthermore, $\mathrm{ML}(A) = k[x]$ and $\mathscr{D}(A) = k[x, z, t]$, where x, z, t are the congruence classes of X, Z, T in A.

In a third paper [56] published in 2014, Gupta extends this result to obtain counterexamples to the Cancellation Problem for all dimensions $n \geq 3$.

Theorem 3.4.9 (Gupta) *Let k be a field and let $f(Z, T) \in k[Z, T]$. Let $m \geq 1$ and let $r_i \geq 2$ for $i = 1, \ldots, m$. Let*

$$A = k[X_1, \ldots, X_m, Y, Z, T]/(X_1^{r_1} \cdots X_m^{r_m} Y - f(Z, T)).$$

If $k[Z, T]/(f) \cong k^{[1]}$, then $A^{[1]} = k[X_1, \ldots, X_m]^{[3]} = k^{[m+3]}$. Moreover, if the characteristic of k is prime and if f is not a variable in $k[Z, T]$, then $A \neq k^{[m+2]}$.

The Cancellation Problem in characteristic zero remains open for $n \geq 3$.

Exercises for Sect. 3.4

1. Let k be an algebraically closed field. Let $A = k[X, Y, Z, T]/(X + X^2 Y + Z^2 + T^3)$, the coordinate ring of the Koras–Russell hypersurface. Prove that $A \not\cong k^{[3]}$. [Begin as described in Sect. 3.1, page 67, and use ideas from the proof of Theorem 3.4.2.]

Notes

The notation ML and the name "Makar-Limanov invariant" have become standard. However, several papers (in particular those of Makar-Limanov) use the notation AK and the terminology **AK invariant**. As of this writing the notation for the Derksen invariant is not set in stone, but the name has become standard. Gupta uses the notation DK. The notation HD is also sometimes used. Exercise 4 in Sect. 3.1 and Exercise 3 in Sect. 3.3 are taken from the 2003 paper of Crachiola and Maubach [15]. This paper seems to provide the debut appearance of the term "Derksen invariant" in a publication.

The notion of higher derivations dates back to a 1937 paper [61] of Helmut Hasse and F.K. (Friedrich Karl) Schmidt. Locally finite iterative higher derivations were explored in several papers, particularly in the 1970s, not long after locally nilpotent derivations gained attention through the works of Jacques Dixmier and Rudolf Rentschler, among others (see [33, 102]). The lecture notes [86] of Miyanishi provide a useful resource and an extensive bibliography on lfihds. Exercise 3 in Sect. 3.2 appears in these notes.

Homogenization for locally nilpotent derivations came from the work of Makar-Limanov (see [78]) but can also be found in an earlier paper [59] of Ofer Hadas. The homogenization technique was extended to prime characteristic by Derksen, Hadas, and Makar-Limanov in [30] where Theorem 3.2.14 appears. Exercise 4 in Sect. 3.2 is also due to Hadas. A characteristic zero formulation of that exercise can be found in [60]. A characteristic zero version of Exercise 5 in Sect. 3.2 is used by Makar-Limanov in [81] to give a streamlined proof that the Koras–Russell hypersurface is not isomorphic to $\mathbb{C}^3$.

The proof of Theorem 3.3.6 is taken from [17]. A different proof using topological methods was given by Rajendra Gurjar in [58].

Danielewski never published Theorem 3.3.7, but his preprint inspired many related results. A classification of surfaces of the form $x^n y = p(z)$ in $\mathbb{C}^3$ (including Danielewski's example) was given by Karl-Heinz Fieseler [44]. Such surfaces, now known as **Danielewski surfaces**, have been researched extensively by many authors (for instance Crachiola [13], Adrien Dubouloz [34–36], Dubouloz and Pierre-Marie Poloni [38], Freudenburg and Lucy Moser-Jauslin [49], Makar-Limanov [79, 80], Moser-Jauslin and Poloni [89], Poloni [100], and Jörn Wilkens [132]). Danielewski has no publications and did not pursue a career in academia. However, in 2009 he participated in a conference at University of Nijmegen where he delivered a lucid lecture on the motivation behind his famous result.

In addition to the papers [54–56] which settle the Cancellation Problem in prime characteristic, Gupta has written an informative survey [57] on her work. This paper elaborates on the connections among several interesting problems involving polynomial rings, including the cancellation problem, Asanuma's dilemma, and the affine fibration problem.

The Jacobian Conjecture: New Equivalences

4

As we have seen in [117], there are various equivalent formulations of the Jacobian Conjecture. The aim of this chapter is to give four new equivalent formulations. The first one is the **Dixmier Conjecture**, which is known to *imply* the Jacobian Conjecture (see, for example, [117], Theorem 10.4.2). However in 2005 Y. Tsuchimoto in [116] showed that the converse also holds. So both conjectures are equivalent. A little later an independent proof of this result was given by Belov-Kanel and Kontsevich in [8]. Inspired by their proof, Adjamagbo and the author in [2] give a third proof of the equivalence; this is the proof we essentially follow in this chapter. In [2] a new conjecture is introduced, the so-called **Poisson Conjecture**, and it is shown that it is also equivalent to the Dixmier Conjecture and the Jacobian Conjecture. In Sect. 4.5 we give two more equivalent conjectures.

The contents of this chapter are arranged as follows: in Sect. 4.1 we recall some properties and definitions concerning exterior forms. For more details, we refer to [53]. In Sect. 4.2 we introduce the Poisson algebra and the Poisson Conjecture, and in the next section we recall the Dixmier Conjecture and study Weyl algebras over commutative rings. In Sect. 4.4 we formulate and prove the first main result of this chapter, namely Theorem 4.4.1. Finally, in Sect. 4.5 we prove a recent result of Lipton and the author in [120], which shows that the Jacobian Conjecture is equivalent to the following statement: for each $n \geq 1$ and almost all prime numbers p, each Keller map $F \in \mathbb{Z}_p[x_1, \ldots, x_n]^n$ has the property that its induced map $\overline{F} : \mathbb{F}_p^n \to \mathbb{F}_p^n$ is not the zero map. Also we show that the Jacobian Conjecture is equivalent to a new conjecture, called the **Unimodular Conjecture**.

© The Author(s), under exclusive license to Springer Nature Switzerland AG 2021
A. van den Essen et al., *Polynomial Automorphisms and the Jacobian Conjecture*,
Frontiers in Mathematics, https://doi.org/10.1007/978-3-030-60535-3_4

4.1   Preliminaries: Exterior Forms

Throughout this section, A is a commutative ring and n is a positive integer such that $n!$ is
a unit in A. Let $E := A^n$ be the free A-module of rank n. If $p \geq 1$, we denote by $L^p E$
the set of A-multilinear mappings from $E^p := E \times \cdots \times E \to A$. An element of $L^p E$ is
called a **multilinear form of degree** p **on** E or a p-**form on** E. If $\alpha \in L^p E$ and $\beta \in L^q E$,
one defines the product $\alpha\beta \in L^{p+q} E$ by the formula

$$\alpha\beta(x_1, \ldots, x_p, x_{p+1}, \ldots, x_{p+q}) = \alpha(x_1, \ldots, x_p)\beta(x_{p+1}, \ldots, x_{p+q})$$

for all $x_i \in E$. An **exterior form** α **on** E **of degree** p is an antisymmetric p-form on
E. The set of such forms is an A-module and will be denoted by $\Lambda^p E$. It follows that
$\Lambda^p E = 0$ if $p > n$. To an element $\alpha \in L^p E$, one associates an element $a(\alpha) \in \Lambda^p E$ by
defining

$$a(\alpha) := \sum_{\sigma \in S_p} \varepsilon(\sigma)\sigma \circ \alpha,$$

where $(\sigma \circ \alpha)(x_1, \ldots, x_p) = \alpha(x_{\sigma^{-1}(1)}, \ldots, x_{\sigma^{-1}(p)})$, for all $x_i \in E$. Furthermore, if
$1 \leq p, q \leq n, \alpha \in \Lambda^p E$, and $\beta \in \Lambda^q E$, the **exterior product of** α **and** β is defined by

$$\alpha \wedge \beta := \frac{1}{p!q!} a(\alpha\beta).$$

This product is associative and satisfies $\alpha \wedge \beta = (-1)^{pq} \beta \wedge \alpha$. If $h : E \to E$ is an A-linear
map and $\alpha \in L^p E$, we obtain an element $h^*\alpha \in L^p E$ by defining

$$(h^*\alpha)(x_1, \ldots, x_p) = \alpha(h(x_1), \ldots, h(x_p))$$

for all $x_i \in E$. Furthermore, if $\alpha \in \Lambda^p E$, then $h^*\alpha \in \Lambda^p E$ and $h^*(\alpha \wedge \beta) = h^*\alpha \wedge h^*\beta$
for all $\beta \in \Lambda^q E$.

Now let $(e) = (e_1, \ldots, e_n)$ be the standard basis of E and let $(e^*) = (e_1^*, \ldots, e_n^*)$
be its dual basis. So each e_i^* belongs to $L^1 E = \Lambda^1 E$. The forms $e_{i_1}^* \wedge \cdots \wedge e_{i_p}^*$, with
$1 \leq i_1 < \cdots < i_p \leq n$, form an A-basis of the free A-module $\Lambda^p E$. So the rank of this
module equals $\binom{n}{p}$. In particular, $\Lambda^n E$ is a free A-module of rank 1. A basis element ω of
$\Lambda^n E$ is called a **volume form**. Since obviously $e_1^* \wedge \cdots \wedge e_n^*$ is a volume form, we get that
$\omega = ae_1^* \wedge \cdots \wedge e_n^*$, for some unit a in A. It follows from Exercise 1 that if $h : E \to E$ is
an A-linear map, then $h^*\omega = (\det h)\omega$.

4.2   The Canonical Poisson Algebra and the Poisson Conjecture

Let R be a commutative ring with $1 \in R$ and n a positive integer. The n**th canonical Poisson algebra** $P_n(R)$ **over** R is the polynomial ring $A := R[x_1, \ldots, x_{2n}]$ equipped with the **canonical Poisson bracket** $\{\,,\}$ defined by

$$\{f, g\} = \sum_{i=1}^{n} \left(\frac{\partial f}{\partial x_i} \frac{\partial g}{\partial x_{i+n}} - \frac{\partial f}{\partial x_{i+n}} \frac{\partial g}{\partial x_i} \right).$$

This Poisson bracket is R-bilinear and antisymmetric and is a derivation in each of its components, i.e., $\{a, bc\} = \{a, b\}c + b\{a, c\}$ for all $a, b, c \in A$. It follows that $\{\,,\}$ is completely determined by the values $\{x_i, x_j\}$.

An R-endomorphism ϕ of A is called an **endomorphism of** $P_n(R)$ if

$$\phi(\{f, g\}) = \{\phi(f), \phi(g)\}$$

for all $f, g \in A$. By Exercise 2, this is equivalent to

$$\{\phi(x_i), \phi(x_j)\} = \{x_i, x_j\}, \text{ for all } i, j.$$

Now we will prove the main result of this section, Theorem 4.2.1 below, which states that if $n!$ is a unit in R and ϕ is an endomorphism of $P_n(R)$, then $det\ JF = 1$, where F is the corresponding polynomial map given by $F = (\phi(x_1), \ldots, \phi(x_{2n}))$. In order to prove this result, we need some preliminaries.

Let $E := A^{2n}$, the free A-module with standard basis $(e) := (e_1, \ldots, e_{2n})$ and $(e^*) := (e_1^*, \ldots, e_{2n}^*)$ its dual basis. The **canonical symplectic form** on E is the bilinear form given by

$$\omega := \sum_{i=1}^{n} e_i^* \wedge e_{i+n}^*.$$

From the definition in Sect. 4.1, we obtain: if $p < q$, then $(e_i^* \wedge e_{i+n}^*)(e_p, e_q) = 1$ if $i = p$ and $i + n = q$ and 0 otherwise. An A-linear endomorphism L of E is called **symplectic** if $L^*\omega = \omega$, and finally a polynomial map $F := (F_1, \ldots, F_{2n}) \in A^{2n}$ is called **symplectic** if the A-linear map on E defined by the matrix $(JF)^t$ is symplectic.

Theorem 4.2.1 *There is equivalence between*

(i) F is symplectic and
(ii) $\phi \in End_R A$ defined by $\phi(x_i) = F_i$ is an endomorphism of $P_n(R)$.

Furthermore, if (i) or (ii) holds and $n!$ is a unit in R, then $det\ JF = 1$.

The proof of this result is based on the following lemma, where for a bilinear form B on E we denote by $M_{(e)}(B)$ the matrix $(B(e_i, e_j))_{1 \le i, j \le 2n}$.

Lemma 4.2.2 *Let $F = (F_1, \ldots, F_{2n}) \in A^{2n}$. Then*

$$M_{(e)}(((JF)^t)^* \omega) = (\{F_i, F_j\})_{1 \le i, j \le 2n}.$$

Proof Let $1 \le p, q \le 2n$. Then $(((JF)^t)^* \omega)(e_p, e_q) = \omega((JF)^t e_p, (JF)^t e_q)$. Observe $(JF)^t e_p = \sum_{j=1}^{2n} \frac{\partial F_p}{\partial x_j} e_j$. So

$$(e_i^* \wedge e_{i+n}^*)((JF)^t e_p, (JF)^t e_q) = \frac{\partial F_p}{\partial x_i} \frac{\partial F_q}{\partial x_{i+n}} - \frac{\partial F_p}{\partial x_{i+n}} \frac{\partial F_q}{\partial x_i},$$

which gives the desired result. $\square$

Corollary 4.2.3 *Let $F = (F_1, \ldots, F_{2n}) \in A^{2n}$. There is equivalence between*

(i) *F is symplectic,*
(ii) *$\{F_i, F_j\} = \{x_i, x_j\}$ for all $1 \le i, j \le 2n$, and*
(iii) *$\phi \in End_R A$ defined by $\phi(x_i) = F_i$ is an endomorphism of $P_n(R)$.*

Proof By taking $F = (x_1, \ldots, x_{2n})$ in the above lemma, we see that we have the equality $M_{(e)}(\omega) = (\{x_i, x_j\})_{1 \le i, j \le 2n}$. Furthermore, since two bilinear forms B_1 and B_2 are equal if and only if their matrices $M_{(e)}(B_1)$ and $M_{(e)}(B_2)$ are equal, the equivalence of (i) and (ii) follows from the lemma above. Finally, the equivalence of (ii) and (iii) follows from the earlier observation that ϕ is an endomorphism of $P_n(R)$ if and only if $\{\phi(x_i), \phi(x_j)\} = \{x_i, x_j\}$ for all i, j. $\square$

Proof of Theorem 4.2.1 The first statement follows from Corollary 4.2.3. So let F be symplectic and let as before L be the A-linear map of $E = A^{2n}$ defined by the matrix $(JF)^t$. Put $v := e_1^* \wedge \cdots \wedge e_{2n}^*$ the standard volume form on E. Since $\omega^n = n!(-1)^{\frac{n(n-1)}{2}} v$ by Exercise 3 and $n!$ is a unit in R, it follows that ω^n is a volume form on E. Furthermore, since F is symplectic, $L^* \omega = \omega$ and hence $L^*(\omega^n) = (L^*(\omega))^n = \omega^n$. On the other hand, since ω^n is a volume form, it follows from the last line of Sect. 4.1 that $L^*(\omega^n) = (det \, L)\omega^n$. Hence $\omega^n = (det \, L)\omega^n$, which implies that $det \, L = 1$. Since $det \, L = det \, (JF)^t = det \, JF$, we get $det \, JF = 1$, as desired. $\square$

Poisson Conjecture *For every $n \ge 1$ and every field k of characteristic zero, the following statement holds:*
$PC(n, k)$ Every endomorphism of $P_n(k)$ is an automorphism.

As an immediate consequence of Theorem 4.2.1, we obtain that the Jacobian Conjecture implies the Poisson Conjecture. More precisely,

Proposition 4.2.4 $JC(2n, k)$ *implies* $P_n(k)$.

4.3   The Weyl Algebra and the Dixmier Conjecture

Throughout this section, R denotes a commutative ring with $1 \in R$ and n is a positive integer. The n**th Weyl algebra over** R, denoted $A_n(R)$, is the associative R-algebra with generators $y_1, \ldots, y_{2n}$ and relations

$$[y_i, y_{i+n}] = 1 \text{ for all } 1 \leq i \leq n \text{ and } [y_i, y_j] = 0 \text{ otherwise.}$$

Dixmier Conjecture *For every* $n \geq 1$ *and every field* k *of characteristic zero, the following statement holds:*
$DC(n, k)$ *Every endomorphism of* $A_n(k)$ *is an automorphism.*

It is well known that the Dixmier Conjecture implies the Jacobian Conjecture, more precisely $DC(n, k)$ implies $JC(n, k)$ (see [117]). Consequently, in order to prove the equivalences between the Dixmier, Jacobian, and Poisson conjectures, it remains to show that the Poisson Conjecture implies the Dixmier Conjecture (for in Sect. 4.2 we showed that the Jacobian Conjecture implies the Poisson Conjecture). In fact we will show in Sect. 4.4 that $P(n, \mathbb{C})$ implies $D(n, \mathbb{C})$ and deduce that the Poisson Conjecture implies the Dixmier Conjecture. To prove these results, we will first study Weyl algebras over R, where R is a ring of finite characteristic. More precisely, from now on *until Lemma 4.3.5* , we will assume that R is nonzero and that *there exists a prime number p such that $p \cdot 1 = 0$*. It follows that

(*)   if $0 \neq r \in R$ and $m \cdot r = 0$, with $m \in \mathbb{Z}$, then $m \in p\mathbb{Z}$.

The following proposition is well known. Let $Z := Z(A_n(R))$ denote the center of $A_n(R)$, i.e., the set of all elements of $A_n(R)$, which commute with all elements of $A_n(R)$.

Proposition 4.3.1 $Z = R[y_1^p, \ldots, y_{2n}^p]$.

Proof Since $[y_i^p, y_j] = 0$ for all $j \neq i + n$ and $[y_i^p, y_{i+n}] = py_i^{p-1} = 0$, we get that $y_i^p \in Z$, whence $R[y_1^p, \ldots, y_{2n}^p] \subseteq Z$. To prove the converse, write $A_n(R) = S[y_n, y_{2n}]$, where $S = A_{n-1}(R)$. Let $z = \sum a_i(y_n) y_{2n}^i \in Z$, with $a_i(y_n) \in S[y_n]$. Then $[y_n, z] = 0$ implies that $\sum i a_i(y_n) y_{2n}^{i-1} = 0$. So by (*), we get

$$z = \sum a_{pj}(y_n) y_{2n}^{pj}.$$

Using $[z, y_{2n}] = 0$, we also find that $a_{pj}(y_n) \in S[y_n^p]$ for all j. Hence $Z \subseteq S[y_n^p, y_{2n}^p]$. Finally, writing z as a polynomial in y_n^p, y_{2n}^p with coefficients in S, it follows by induction on n, using that $[z, y_i] = [z, y_{i+n}] = 0$ for all $1 \leq i \leq n-1$, that z belongs to $R[y_1^p, \ldots, y_{2n}^p]$, which completes the proof.   □

Now let ϕ be an R-endomorphism of $A_n(R)$. Put $P_i = \phi(y_i)$. The main result of this section is

Theorem 4.3.2 $A_n(R) = \bigoplus_{0 \leq \alpha_i < p} Z P_1^{\alpha_1} \cdots P_{2n}^{\alpha_{2n}}$.

Before we prove this result, we deduce some fundamental consequences.

Proposition 4.3.3

(i) $\phi(Z) \subseteq Z$.
(ii) If $\phi(Z) = Z$, then $\phi(A_n(R)) = A_n(R)$.
(iii) If $\phi_{|Z} : Z \to Z$ is injective, then $\phi : A_n(R) \to A_n(R)$ is injective.

Proof

(i) Let $z \in Z$. Then $\phi(z)$ commutes with each $\phi(y_i) = P_i$, hence with each element $P_1^{\alpha_1} \cdots P_{2n}^{\alpha_{2n}}$. Also $\phi(z)$ commutes with Z. So by Theorem 4.3.2, $\phi(z)$ commutes with $A_n(R)$, i.e., $\phi(z) \in Z$.
(ii) Let $a \in A_n(R)$. By Theorem 4.3.2, we can write a in the form

$$a = \sum z_\alpha P_1^{\alpha_1} \cdots P_{2n}^{\alpha_{2n}}, \ z_\alpha \in Z.$$

Since $\phi(Z) = Z$, there exists $z_\alpha^* \in Z$ such that $z_\alpha = \phi(z_\alpha^*)$. Also $P_i = \phi(y_i)$. Hence

$$a = \phi(\sum z_\alpha^* y_1^{\alpha_1} \cdots y_{2n}^{\alpha_{2n}}) \in \phi(A_n(R)).$$

So $A_n(R) = \phi(A_n(R))$.
(iii) Let $a \in A_n(R)$ with $\phi(a) = 0$. Observe that by Theorem 4.3.2, a can be written in the form $a = \sum z_\alpha y_1^{\alpha_1} \cdots y_{2n}^{\alpha_{2n}}$, with $z_\alpha \in Z$ and all $0 \leq \alpha_i < p$. Then $\phi(a) = 0$ implies that $\sum \phi(z_\alpha) P_1^{\alpha_1} \cdots P_{2n}^{\alpha_{2n}} = 0$. Since by (i) each $\phi(z_\alpha)$ belongs to Z, it follows from the direct sum decomposition in Theorem 4.3.2 that all $\phi(z_\alpha)$ are zero. From the injectivity of $\phi_{|Z}$, it then follows that all z_α are zero, which gives that a is zero.   □

Proof of Theorem 4.3.2 (Started) Consider the Z-module

$$N = \sum_{0 \leq \alpha_i < p} Z P_1^{\alpha_1} \cdots P_{2n}^{\alpha_{2n}}.$$

Using the inner derivations $[P_i, -]$ on $A_n(R)$, one readily obtains that N is a free Z-module of rank p^{2n}. From 4.3.1, we obtain that

$$M := A_n(R) = \sum_{0 \leq \alpha_i < p} Z y_1^{\alpha_1} \cdots y_{2n}^{\alpha_{2n}},$$

and using the derivations $[y_i, -]$, it follows that M is also a free Z-module of rank p^{2n}.

First assume that R is a *domain* and put $S = Z \backslash \{0\}$. Since $N \subseteq M$, we get that $S^{-1}N \subseteq S^{-1}M$. Furthermore, both modules are $S^{-1}Z$-vector spaces of dimension p^{2n}, hence they are equal. So there exists a nonzero element u in Z with $uM \subseteq N$, i.e., for each $m \in M$, there exist $u_\alpha \in Z$ such that

$$um = \sum_{0 \leq \alpha_i < p} u_\alpha P_1^{\alpha_1} \cdots P_{2n}^{\alpha_{2n}}.$$

From Lemma 4.3.4 below, we obtain that $u | u_\alpha$ in Z for each α. Since $A_n(R)$ is a domain, it follows that $m \in N$, whence $M \subseteq N$. So $A_n(R) = N$, as desired. $\square$

Lemma 4.3.4 *If $m \in M$ satisfies*

$$um = \sum_{0 \leq \alpha_i < p} u_\alpha P_1^{\alpha_1} \cdots P_{2n}^{\alpha_{2n}}, \tag{4.3.1}$$

then $u | u_\alpha$ in Z for each α.

Proof Let $m \neq 0$ and use induction on $d := \max\{|\alpha|, |u_\alpha| \neq 0\}$. If $d = 0$, then $um = u_0$, $u_0 \in Z$. Write $m = \sum z_\alpha y^\alpha$ on the free Z-basis y^α, $0 \leq \alpha_i < p$. Then $u_0 = um = \sum(uz_\alpha)y^\alpha$. So $u_0 = uz_0$, i.e., $u | u_0$ in Z.

Now let $d > 0$ and $u_\alpha \neq 0$, with $\alpha_j > 0$. Applying the inner derivation $[P_j, -]$ to the Eq. (4.3.1) gives

$$u[P_j, m] = \sum u_\alpha \alpha_{2j} P_1^{\alpha_1} \cdots P_{2j}^{\alpha_{2j}-1} \cdots P_{2n}^{\alpha_{2n}}.$$

Observe that $0 < \alpha_{2j} < p$ and $u_\alpha \neq 0$, so by (*) $u_\alpha\alpha_{2j} \neq 0$. The induction hypothesis implies that $u|\alpha_{2j}u_\alpha$, whence $u|u_\alpha$ in Z. Repeating this argument, we obtain that $u|u_\alpha$ for all α with $|\alpha| > 0$ and $u_\alpha \neq 0$. Say $u_\alpha = uv_\alpha$, $v_\alpha \in Z$. Then

$$um = u_0 + \sum_{|\alpha|>0, 0\leq\alpha_i<p} uv_\alpha P^\alpha.$$

So $u(m - \sum v_\alpha P^\alpha) = u_0$. Hence by the case $d = 0$ above, it follows that $u|u_0$ in Z. So $u|u_\alpha$ in Z for all α, as desired. □

Proof of Theorem 3.2 (Finished)

(i) Let R be an arbitrary commutative ring. Replacing R by the finitely generated $\mathbb{Z}$-algebra of R generated by the coefficients of the P_i with respect to the monomials y^α, we may assume that R is a finitely generated $\mathbb{Z}$-algebra. In particular, R is noetherian and hence its nilradical is a finite intersection of prime ideals. Since a finite power of the nilradical equals the zero ideal, we obtain that the zero ideal is a finite product of prime ideals, say $(0) = \mathfrak{p}_1\mathfrak{p}_2\cdots\mathfrak{p}_s$.

(ii) Let $\mathfrak{p}$ be a prime ideal in R. Then $\overline{R} := R/\mathfrak{p}$ is a domain. So, as shown above

$$A_n(\overline{R}) \subseteq \sum_{0\leq\alpha_i<p} Z(A_n(\overline{R}))\overline{P}_1^{\alpha_1}\cdots\overline{P}_{2n}^{\alpha_{2n}}$$

$$= \sum_{0\leq\alpha_i<p} \overline{R}[y_1^p,\ldots,y_{2n}^p]\overline{P}_1^{\alpha_1}\cdots\overline{P}_{2n}^{\alpha_{2n}}.$$

Hence

$$A_n(R) \subseteq \sum_{0\leq\alpha_i<p} R[y_1^p,\ldots,y_{2n}^p]P_1^{\alpha_1}\cdots P_{2n}^{\alpha_{2n}} + \mathfrak{p}A_n(R).$$

So by Proposition 4.3.1, we get $A_n(R) \subseteq N + \mathfrak{p}A_n(R)$.

(iii) Finally, we use that $\mathfrak{p}_1\mathfrak{p}_2\cdots\mathfrak{p}_s = (0)$ and that $\mathfrak{p}_i \subseteq R \subseteq Z$: by (ii) we get that $A_n(R) \subseteq N + \mathfrak{p}_1 A_n(R)$ and $A_n(R) \subseteq N + \mathfrak{p}_2 A_n(R)$, whence $A_n(R) \subseteq N + \mathfrak{p}_1 N + \mathfrak{p}_1\mathfrak{p}_2 A_n(R) \subseteq N + \mathfrak{p}_1\mathfrak{p}_2 A_n(R)$. Repeating this argument gives

$$A_n(R) \subseteq N + \mathfrak{p}_1\cdots\mathfrak{p}_s A_n(R) = N.$$

So $A_n(R) = N$, as desired. □

Now let ϕ be an R-endomorphism of $A_n(R)$. The degree of ϕ, denoted by $\deg_y\phi$, is by definition the maximum of the degrees of the elements $\phi(y_i)$, where the degree of a nonzero element a of $A_n(R)$ is the maximum of the degrees of the monomials cy^α appearing in a. By Proposition 4.3.1 and Proposition 4.3.3 (i), it follows that ϕ induces an R-endomorphism of the polynomial ring $Z(A_n(R)) = R[y_1^p, \ldots, y_{2n}^p]$, which we denote by ϕ_{pol}.

Writing x_i instead of y_i^p for all i, the degree of ϕ_{pol}, denoted by $\deg_x \phi_{pol}$, is the maximum of all degrees of the polynomials $\phi_{pol}(x_i)$ in $R[x_1, \ldots, x_{2n}]$. In the next section, the following result will be useful.

Lemma 4.3.5 *If R is a domain, then $\deg_y \phi = \deg_x \phi_{pol}$.*

Proof Since $\phi(y_i^p) \in R[y_1^p, \ldots, y_{2n}^p]$, we get that

$$\phi(y_i^p) = \sum c_j (y_1^p)^{j_1} \cdots (y_{2n}^p)^{j_{2n}}$$

for some $c_j \in R$. So

$$\phi_{pol}(x_i) = \sum c_j x_1^{j_1} \cdots x_{2n}^{j_{2n}},$$

whence $\deg_x \phi_{pol}(x_i) = \max_{c_j \neq 0}(j_1 + \cdots + j_{2n})$. Consequently $\deg_y \phi(y_i^p) = p \deg_x \phi_{pol}(x_i)$. Finally, since $\deg_y \phi(y_i^p) = \deg_y \phi(y_i)^p = p \deg_y \phi(y_i)$, for R is a domain, the desired result follows. $\qquad\square$

Applications to the Case that R is a Domain

In the remainder of this section, we will assume that R is a domain of characteristic zero and $\mathfrak{m}$ is a maximal ideal of R such that $R/\mathfrak{m}$ is a finite field. So there exists a prime number p such that $p \cdot \bar{1} = 0$, i.e. $p \in \mathfrak{m}$.

Let ϕ be an R-endomorphism of $A_n(R)$. It induces an R/pR-endomorphism $\bar\phi$ of $A_n(R/pR)$, which, as observed above, induces an R/pR-endomorphism $\bar\phi_{pol}$ of the polynomial ring $R/pR[x_1, \ldots, x_{2n}]$.

Theorem 4.3.6 $\bar\phi_{pol}$ *is an endomorphism of* $P_n(R/pR)$.

To prove this theorem, we give another description of the Poisson bracket on $P_n(R/pR)$. Put $W := R[y_1^p, \ldots, y_{2n}{}^p]$ and $x_i := y_i^p$ for each i.

Lemma 4.3.7 *Let $A \in W$ and $B \in A_n(R)$. Then $[A, B] \in pA_n(R)$.*

Proof Since $[\,,\,]$ is R-bilinear, we may assume that $A = (y_1^p)^{\alpha_1} \cdots (y_{2n}^p)^{\alpha_{2n}}$ and $B = y_1^{\beta_1} \cdots y_{2n}^{\beta_{2n}}$. Using Leibniz' rule, we may even assume that $A = y_i^p$ and $B = y_j$, in which case the result is clear. $\square$

Proposition 4.3.8 *Let* $a, b \in R/pR[x_1, \ldots, x_{2n}]$ *and* $A, B \in W$ *such that* $a = A(mod\ pA_n(R))$ *and* $b = B(mod\ pA_n(R))$. *Then* $\frac{1}{p}[A, B]$ *is a well-defined element of* $A_n(R)$ *and*

$$\{a, b\} = \frac{1}{p}[A, B](mod\ pA_n(R)).$$

Proof Since R has no $\mathbb{Z}$-torsion, the first statement follows from Lemma 4.3.7. To prove the formula, observe that both $\{\,,\,\}$ and $[\,,\,]$ are bilinear and antisymmetric and satisfy Leibniz's rule. Therefore, it suffices to show that

$$\{x_i, x_j\} = \frac{1}{p}[y_i^p, y_j^p](mod\ pA_n(R))$$

for all $i < j$. If $j \neq i + n$, both sides are zero. So assume $j = i + n$. Then the result follows from the following formula:

$$\frac{1}{p}[y_{i+n}^p, y_i^p] = \frac{1}{p}\sum_{k=0}^{p-1} \frac{(p!)^2}{(k!)^2(p-k)!} y_i^k y_{i+n}^k = -1\big(mod\ p\mathbb{Z}[y_i, y_{i+n}]\big), \qquad (4.3.2)$$

which can be found in [33]. See also Exercise 4. $\square$

Proof of Theorem 4.3.6 To prove that $\overline{\phi}_{pol}$ is an endomorphism of $P_n(R/pR)$, it suffices to show that $\{\overline{\phi}_{pol}(x_i), \overline{\phi}_{pol}(x_j)\} = \{x_i, x_j\}$ for all $i < j$. Since $y_i^p + pA_n(R) \in Z(A_n(R/pR))$, it follows that $\phi(y_i^p) = A + pA'$ with A in W and A' in $A_n(R)$. Similarly $\phi(y_j^p) = B + pB'$ with B in W and B' in $A_n(R)$. Then by Proposition 4.3.8,

$$\{\overline{\phi}_{pol}(x_i), \overline{\phi}_{pol}(x_j)\} = \frac{1}{p}[A, B](mod\ pA_n(R)).$$

Since by Lemma 4.3.7 both $[A, B']$ and $[A', B]$ belong to $pA_n(R)$, we get

$$\frac{1}{p}[\phi(y_i^p), \phi(y_j^p)] = \frac{1}{p}[A, B]+[A, B']+[A', B]+p[A', B'] = \frac{1}{p}[A, B](mod\ pA_n(R)).$$

So $\{\overline{\phi}_{pol}(x_i), \overline{\phi}_{pol}(x_j)\} = \frac{1}{p}[\phi(y_i^p), \phi(y_j^p)](mod\ pA_n(R))$.

If $j \neq i + n$ the right-hand side is equal to zero and hence equals $\{x_i, x_j\}$. Finally, if $j = i + n$ the right-hand side equals $\phi(\frac{1}{p}[y_i^p, y_{i+n}^p])(mod\ pA_n(R))$, which by (4.3.2) is equal to 1 and hence equals $\{x_i, x_{i+n}\}$, which concludes the proof. $\square$

Let as before ϕ be an R-endomorphism of $A_n(R)$. Reducing modulo $\mathfrak{m}$, we get an $\overline{R} := R/\mathfrak{m}$ endomorphism ψ of $A_n(\overline{R})$, which, since $p \in \mathfrak{m}$, induces an $\overline{R}$-endomorphism ψ_{pol} of the polynomial ring $\overline{R}[x_1, \ldots, x_{2n}]$. Since $pR \subseteq \mathfrak{m}$, we get a surjective ring homomorphism from R/pR to $R/\mathfrak{m} = \overline{R}$ and hence a surjective ring homomorphism from $R/pR[x_1, \ldots, x_{2n}]$ to $\overline{R}[x_1, \ldots, x_{2n}]$. Since $\overline{\phi}_{pol}$ is an endomorphism of $P_n(R/pR)$, by Theorem 4.3.6, it follows that ψ_{pol} is an endomorphism of $P_n(\overline{R})$. Summarizing we have:

Corollary 4.3.9 *If ϕ is an R-endomorphism of $A_n(R)$, then ψ_{pol}, the $\overline{R}$-endomorphism of $\overline{R}[x_1, \ldots, x_{2n}]$ induced by ϕ, is an endomorphism of $P_n(\overline{R})$.*

4.4   The Equivalence of the Dixmier, Jacobian, and Poisson Conjectures

The main result of this section is:

Theorem 4.4.1 *For each $n \geq 1$, we have the following implications:*
$DC(n, \mathbb{C}) \to JC(n, \mathbb{C})$, $JC(2n, \mathbb{C}) \to P(n, \mathbb{C})$ and $P(n, \mathbb{C}) \to DC(n, \mathbb{C})$,
i.e., the Dixmier, Jacobian, and Poisson conjectures are equivalent.

As observed in the beginning of Sect. 4.3, it remains to prove the last implication. First, we study the statement $DC(n, \mathbb{C})$, i.e., the statement that every $\mathbb{C}$-endomorphism of $A_n(\mathbb{C})$ is an automorphism. Since $A_n(\mathbb{C})$ is a simple ring and the kernel of a $\mathbb{C}$-endomorphism of $A_n(\mathbb{C})$ is a two-sided ideal that is not the whole ring, it follows that every $\mathbb{C}$-endomorphism of $A_n(\mathbb{C})$ is injective. Hence $DC(n, \mathbb{C})$ is equivalent to saying that every $\mathbb{C}$-endomorphism of $A_n(\mathbb{C})$ is surjective.

We will show below that $P(n, \mathbb{C})$ implies a more refined statement. To describe it, observe that a $\mathbb{C}$-endomorphism ϕ of $A_n(\mathbb{C})$ is surjective if and only if there exist $\psi_1, \ldots, \psi_{2n}$ in $A_n(\mathbb{C})$ such that $\phi(\psi_i) = y_i$ for all i. If all ψ_i have degree $\leq N$, for some N, we say that ϕ is **N-surjective**.

Proposition 4.4.2 *Let $d \geq 1$. Then $P(n, \mathbb{C})$ implies:*
$DC(n, \mathbb{C}, d)$ *Every $\mathbb{C}$-endomorphism of $A_n(\mathbb{C})$ of degree $\leq d$ is $D := d^{2n-1}$-surjective.*

Proof

(i) By contradiction. So let ϕ be a $\mathbb{C}$-endomorphism of $A_n(\mathbb{C})$, say of degree d, which is not D-surjective. This means that there do *not* exist $\psi_1, \ldots, \psi_{2n}$ of degree $\leq D$ in $A_n(\mathbb{C})$ such that

$$\phi(\psi_1) - y_1 = 0, \ldots, \phi(\psi_{2n}) - y_{2n} = 0.$$

So if we consider the **universal Weyl algebra elements** of degree D, i.e., the expressions $\psi_i^U = \sum_{|\alpha| \leq D} c_\alpha^{(i)} y^\alpha$, where all $c_\alpha^{(i)}$ are different variables and consider the formal expressions

$$\phi(\psi_i^U) - y_i := \sum_{|\alpha| \leq D} c_\alpha^{(i)} \phi(y_1)^{\alpha_1} \ldots \phi(y_{2n})^{\alpha_{2n}} - y_i,$$

then the coefficients $P_1(C), \ldots, P_s(C)$ of all monomials y^α appearing in these expressions (each $P_i(C)$ is a polynomial in the polynomial ring $\mathbb{C}[C]$, which is generated over $\mathbb{C}$ by all variables $c_\alpha^{(i)}$), have no common zero in $\mathbb{C}^C$. So by the Nullstellensatz, there exist $Q_1(C), \ldots, Q_s(C)$ in $\mathbb{C}[C]$ such that

$$1 = \sum Q_j(C) P_j(C). \tag{4.4.1}$$

(ii) By the hypothesis, we know that $P(n, \mathbb{C})$ holds. Consequently, if $d \geq 1$, then the Poisson Conjecture holds for all endomorphisms of $P_n(\mathbb{C})$ of degree $\leq d$. By van den Essen [117] Proposition 2.3.1, this implies that the following more precise statement holds:

$P(n, \mathbb{C}, d)$: if $F = (F_1, \ldots, F_{2n})$ satisfies $F(0) = 0$, $\deg F_i \leq d$ for all i and $\{F_i, F_j\} = \{x_i, x_j\}$ for all i, j, then F has an inverse of degree $\leq d^{2n-1}$.

To rewrite this statement in terms of polynomial equations, we consider the universal polynomial map of degree d in $2n$ variables $x = (x_1, \ldots, x_{2n})$, i.e.,

$$F_i^U := (F_1^U, \ldots, F_{2n}^U),$$

where each $F_i^U := \sum A_\alpha^{(i)} x^\alpha$ with $|\alpha| \leq d$ and all $A_\alpha^{(i)}$ are different variables. Let $A := (\ldots, A_\alpha^{(i)}, \ldots)$ and denote by $\mathbb{Z}[A]$ the polynomial ring in A over $\mathbb{Z}$. Then all the polynomials $P_{ij} := \{F_i^U, F_j^U\} - \{x_i, x_j\}$ for all $1 \leq i < j \leq 2n$ belong to $\mathbb{Z}[A][x]$. Now we replace $\mathbb{Z}$ by $\mathbb{Z}[\frac{1}{n!}]$ and let $J := (g_1(A), \ldots, g_r(A))$ be the ideal in $\mathbb{Z}[\frac{1}{n!}][A]$ generated by the coefficients of all monomials x^α appearing in all P_{ij}. Then by Theorem 4.2.1 and Corollary 4.2.3, the canonical image of F^U in $(\mathbb{Z}[\frac{1}{n!}][A]/J)[x]^{2n}$ has Jacobian determinant 1 and hence, by the formal inverse function theorem, it has a formal inverse in $(\mathbb{Z}[\frac{1}{n!}][A]/J)[[x]]^{2n}$, represented by some $G(A)$ in $\mathbb{Z}[\frac{1}{n!}][A][[x]]^{2n}$. Let I be the ideal in $\mathbb{Z}[\frac{1}{n!}][A]$ generated by the coefficients in $G(A)$ of all x^α with $|\alpha| > D := d^{2n-1}$ and let $h_1(A), \ldots, h_t(A)$ be a system of generators of I.

Since $P(n, \mathbb{C}, d)$ holds, it follows that if $a \in \mathbb{C}^A$ is a zero of J, then $F^U(A = a)$ is invertible with inverse $G(A = a)$. By [117], Proposition 2.3.1 it follows that $\deg G(A = a) \leq d^{2n-1} = D$. Hence a is a zero of I. So every zero of J is a zero of I. Then the Nullstellensatz implies that there exist $b_j^{(i)}$ in $\mathbb{C}[A]$ and a positive integer ρ such that

$$h_i(A)^\rho = \sum_j b_j^{(i)}(A) g_j(A) \quad \text{for all } 1 \leq i \leq t. \tag{4.4.2}$$

(iii) Now let R be the $\mathbb{Z}$-subalgebra of $\mathbb{C}$ generated by $\frac{1}{n!}$, all coefficients of the monomials y^α appearing in the $\phi(y_i)$, all coefficients appearing in the $b_j^{(i)}$, and all coefficients appearing in all Q_j and P_j. Then ϕ is an R-endomorphism of $A_n(R)$, and R is a finitely generated $\mathbb{Z}$-algebra contained in $\mathbb{C}$. Let m be a maximal ideal in R. Then by [9], section 3, no. 4, theorem 3, $\overline{R} := R/\mathfrak{m}$ is a finite field, say of characteristic $p > 0$. So $p \in \mathfrak{m}$. Reducing the equations in (4.4.1) modulo m, we deduce that the endomorphism ψ of $A_n(\overline{R})$, obtained by reducing the coefficients of ϕ mod m, is *not* D-surjective.

(iv) On the other hand by Proposition 4.3.3 (i) the endomorphism ψ of $A_n(\overline{R})$ induces an endomorphism ψ_{pol} on the polynomial ring $Z(A_n(\overline{R}))$. From Corollary 4.3.9 and Lemma 4.3.5, we deduce that ψ_{pol} is an endomorphism of $P_n(\overline{R})$ of degree $\leq d$. It then follows from (4.4.2) that ψ_{pol} has an inverse of degree $\leq D = d^{2n-1}$. (We may assume that each polynomial $\psi_{pol}(x_i)$ has no constant term.) It follows from Proposition 4.3.3 (ii), (iii) that ψ is an automorphism of $A_n(\overline{R})$. Let τ be its inverse. Then the restriction of τ to the center of $A_n(\overline{R})$, denoted τ_{pol}, is equal to $\overline{\phi}_{pol}^{-1}$ and hence, as observed before, has degree $\leq D$. Since $\deg_x \tau_{pol} = \deg_y \tau$ (by Lemma 4.3.5), we deduce that τ has degree $\leq D$. So ψ is D-surjective, contradicting (iii). $\square$

Theorem 4.4.3 *The Dixmier, Poisson, and Jacobian conjectures are equivalent.*

Proof It remains to show that the Poisson Conjecture implies the Dixmier Conjecture. So let $n \geq 1$ and k be a field of characteristic zero. Let $d \geq 1$ and ϕ a k-endomorphism of $A_n(k)$ of degree $\leq d$. Since by our hypothesis $P(n, \mathbb{C})$ holds, it follows from 4.4.2 that $DC(n, \mathbb{C}, d)$ holds. So there exist $\psi_1, \ldots, \psi_{2n}$ of degree $\leq D := d^{2n-1}$ in $A_n(\mathbb{C})$ such that

$$\phi(\psi_1) - y_1 = 0, \ldots, \phi(\psi_{2n}) - y_{2n} = 0.$$

So if we consider the **universal Weyl algebra elements** of degree D, i.e., the expressions $\psi_i^U = \sum_{|\alpha| \leq D} c_\alpha^{(i)} y^\alpha$, where all $c_\alpha^{(i)}$ are different variables and consider the formal expressions

$$\phi(\psi_i^U) - y_i := \sum_{|\alpha| \leq D} c_\alpha^{(i)} \phi(y_1)^{\alpha_1} \cdots \phi(y_{2n})^{\alpha_{2n}} - y_i,$$

then the coefficients $P_1(C), \ldots, P_s(C)$ of all monomials y^α appearing in these expressions have a common zero in $\mathbb{C}^C$. Observe that all these polynomials are linear in the c_α^i and have coefficients in k. So by a well-known result from linear algebra, it follows that these linear equations also have a solution in k^C, which implies that $D(n, k, d)$ holds. This concludes the proof. $\square$

Remark 4.4.4 It was observed by Alain Kraus and communicated to the first author by Adjamagbo that the proof of the equivalence of the Dixmier and Jacobian Conjectures as given by Belov-Kanel and Kontsevich is based on the following false statement: "if R is a finitely generated $\mathbb{Z}$-algebra contained in $\mathbb{C}$, then for all prime numbers p, except possibly finitely many of them, the ring R/pR is either zero or a domain." A counterexample to this statement is the ring $R = \mathbb{Z}[e^{\pi i/8}]$. In this case for every p, the ring R/pR is nonzero and not a domain.

4.5   A p-Adic Formulation of the Jacobian Conjecture and the Unimodular Conjecture

Throughout this section, R is a commutative ring containing 1 and $n \geq 1$ is a positive integer. If $a = (a_1, \ldots, a_n) \in R^n$, we denote by (a) the ideal in R generated by the a_i, i.e., $(a) = \sum_i Ra_i$. If $(a) = R$, then a is called **unimodular**.

Unimodular Conjecture *For every $n \geq 1$ and every commutative ring R contained in a $\mathbb{Q}$-algebra, the following statement holds:*
$UC(R, n)$ If $F \in R[x]^n$ with $\det JF = 1$, then $F(b)$ is unimodular for some $b \in R^n$.

Proposition 4.5.1 *The Jacobian Conjecture implies the Unimodular Conjecture.*

Proof Let $F \in R[x]^n$ with $\det JF = 1$, where R is a commutative ring contained in a $\mathbb{Q}$-algebra. Since the Jacobian Conjecture over $\mathbb{C}$ implies the Jacobian Conjecture over all such rings R ([117], Proposition 1.1.12), F has a polynomial inverse, say G, over R. Put $b := G(e_1)$, where $e_1 = (1, 0, \ldots, 0)$. Then $F(b) = e_1$ is unimodular. □

In the remainder of this section, we will show that the converse also holds, in other words, that the Unimodular Conjecture and the Jacobian Conjecture are equivalent. To prove this, we will show that if the Unimodular Conjecture holds for the ring of p-adic integers $\mathbb{Z}_p$ for sufficiently many primes p, then the Jacobian Conjecture follows. Furthermore, we will deduce one more surprising equivalent formulation of the Jacobian Conjecture (Theorem 4.5.10). First, we need some preparations.

Transitivity

In this subsection, we investigate 2-**transitivity of** $\mathrm{Aut}_R R^{[n]}$ **on** R^n, i.e., we investigate under which conditions on R any two points $a \neq b$ of R^n can be sent to any two points $c \neq d$ of R^n by an automorphism of $R^{[n]}$.

Let a, b be two different elements of R^n and c, d another such pair. A **morphism** from $V = \{a, b\}$ to $W = \{c, d\}$ is a polynomial map $F = (f_1, \ldots, f_n) \in R[X]^n$ such that

$F(a) = c$ and $F(b) = d$. We say that V and W are **isomorphic** if there exists a morphism F from V to W and a morphism G from W to V such that $G \circ F = 1_V$ and $F \circ G = 1_W$. Our first aim is to investigate under what conditions two sets V and W are isomorphic.

Proposition 4.5.2

(i) *There exists a morphism $V \to W$ if and only if $(d - c) \subseteq (b - a)$.*
(ii) *V and W are isomorphic if and only if $(d - c) = (b - a)$.*

Proof (i) ($\Rightarrow$) Let F be a morphism sending a to c and b to d. Then $G := (x - c) \circ F$ is a morphism sending a to 0 and b to $d - c$. Let $G = (g_1, \ldots, g_n)$. Then $g_i(a) = 0$ implies that $g_i = p_{i1}(x)(x_1 - a_1) + \cdots + p_{in}(x)(x_n - a_n)$, for some $p_{ij}(x)$ in $R[x]$. Since $G(b) = d - c$, we deduce that

$$d_i - c_i = p_{i1}(b)(b_1 - a_1) + \cdots + p_{in}(b)(b_n - a_n) \text{ for all } i.$$

($\Leftarrow$) Since $(d - c) \subseteq (b - a)$, there exist $p_{ij} \in R$ such that

$$d_i - c_i = p_{i1}(b_1 - a_1) + \cdots + p_{in}(b_n - a_n) \text{ for all } i.$$

Let $G = (g_1, \ldots, g_n)$, where $g_i = p_{i1}(x_1 - a_1) + \cdots + p_{in}(x_n - a_n)$. Then $G(a) = 0$ and $G(b) = d - c$. Now put $F = (x + c) \circ G$. Then $F(a) = c$ and $F(b) = d$. So F is a morphism from V to W. This proves (i). Finally, (ii) follows readily from (i), which completes the proof. □

In the next theorem, we assume that R is a PID. We will show that in case V and W are isomorphic, the isomorphism can be extended to an automorphism of R^n, i.e., there exists an $F \in Aut_R R^{[n]}$ such that $F(a) = c$ and $F(b) = d$.

Theorem 4.5.3 *If $\{a, b\}$ and $\{c, d\}$ are isomorphic, then there exists an affine automorphism f of $R^{[n]}$ with $\det Jf = 1$ such that $f(a) = c$ and $f(b) = d$.*

Proof Since R is a PID, there exists $g \in R$ such that $(b - a) = Rg$. Write $b_i - a_i = gv_i$, for some $v_i \in R$. Then $v := (v_1, \ldots, v_n)$ is a unimodular row. Since R is a PID, this implies that there exists a matrix $B \in SL_n(R)$ which first column equals v^t. Let A be the inverse of B and $(r_{i1}, r_{i2}, \ldots, r_{in})$ denote the ith row of A. Define

$$F_i := r_{i1}(x_1 - a_1) + \cdots + r_{in}(x_n - a_n) \text{ for all } 1 \le i \le n.$$

Then, $F = (F_1, \ldots, F_n)$ satisfies $F(a) = 0$. Now we compute $F(b)$. Let $1 \le i \le n$. Then

$$F_i(b) = r_{i1}(b_1 - a_1) + \cdots + r_{in}(b_n - a_n) = g(r_{i1}v_1 + \cdots + r_{in}v_n).$$

But this element is exactly g times the product of the ith row of A and the first column of B. Since $AB = I_n$, this product equals 0 if $i > 1$, i.e., $F_i(b) = 0$ if $i > 1$, and the product equals g if $i = 1$, i.e., $F_1(b) = g$. Clearly, F is an affine automorphism with det $JF = 1$ sending a to 0 and b to ge_1, where e_1 is the first unit standard basis vector. Since by Proposition 4.5.2 $(d - c) = (b - a) = Rg$, we can apply the same argument to find an affine automorphism G with det $JG = 1$ such that $G(c) = 0$ and $G(d) = ge_1$. Then one readily verifies that $f := G^{-1} \circ F$ is an affine automorphism with det $Jf = 1$ such that $f(a) = c$ and $f(b) = d$.   □

Keller Maps and the Unimodular Conjecture

Recall that a polynomial map $F \in R[x]^n$ is called a *Keller map* if $det\ JF = 1$. Below we will use the unimodular conjecture in the following form:

Proposition 4.5.4 *Assume that the unimodular conjecture holds for R, then for every Keller map F and every $a \in R^n$, there exists $d \in R^n$ such that $F(d) - F(a)$ is unimodular.*

Proof Put $G(x) = F(x + a) - F(a)$. Then G is a Keller map. So by the unimodular conjecture, there exists $b \in R^n$ such that $G(b)$ is unimodular, i.e., such that $F(b+a) - F(a)$ is unimodular. Then take $d = b + a$.   □

Theorem 4.5.5 *Let R be a PID and assume that the unimodular conjecture holds for R. If there exists a Keller map such that $F : R^n \to R^n$ is not injective, then for every $m \geq 2$ there exists a Keller map that has a fiber containing at least m elements.*

Proof

   (i) It suffices to show that if F is a Keller map such that $F(a_1) = \cdots = F(a_m) = c$, where $m \geq 2$ and all a_i are different, then there exists a Keller map G such that $\#G^{-1}(c) \geq m + 1$.
   (ii) Since $F(a_1) = F(a_2)$, it follows from [11] or [117], Lemma 10.3.11 ii), that $(a_2 - a_1) = R$. By Proposition 4.5.4, there exists an element d such that $(F(d) - F(a_1)) = R$. So $(F(d) - F(a_1)) = (a_2 - a_1)$. By Proposition 4.5.2, using that $c = F(a_1)$, this means that $\{F(d), c\}$ is isomorphic to $\{a_2, a_1\}$. By Theorem 4.5.3, this implies that there exists a Keller map T such that $T(F(d)) = a_2$ and $T(c) = a_1$.
   (iii) Now put $G = F \circ T \circ F$. Then clearly G is a Keller map. Furthermore,

$$G(a_i) = F \circ T(F(a_i)) = F(T(c)) = F(a_1) = c \text{ for all } 1 \leq i \leq m$$

and

$$G(d) = F \circ T(F(d)) = F(T(F(d))) = F(a_2) = c.$$

Finally, observe that d is different from all a_i, since $F(a_i) = c$ for each i and $(F(d) - F(a_1)) = R$. So $G^{-1}(c)$ contains at least $m + 1$ elements.   □

The Unimodular Conjecture over $\mathbb{Z}_p$

Now we will show that if the unimodular conjecture holds for the ring of p-adic integers for infinitely many primes p, then the Jacobian Conjecture holds. This result is based on the following classical result, which is a special case of a version of Hensel's lemma ([9], Chap. III, section 4, Corollaire 2):

Theorem 4.5.6 (Hensel) *Let $F \in \mathbb{Z}_p[x]^n$ be a Keller map. If $a \in \mathbb{Z}_p^n$ is such that $F(a)$ is in $(p\mathbb{Z}_p)^n$, then there exists a unique $b \in \mathbb{Z}_p^n$ such that $F(b) = 0$ and $b_i \equiv a_i \pmod{p\mathbb{Z}_p}$ for all i.*

Theorem 4.5.7 *If $F \in \mathbb{Z}_p[x]^n$ is a Keller map and $c \in \mathbb{Z}_p^n$, then $\#F^{-1}(c) \leq p^n$.*

Proof If $\#F^{-1}(c) = 0$ we are done, so assume that $\#F^{-1}(c) \geq 1$, say $c = F(a)$ for some $a \in \mathbb{Z}_p^n$. Then $G = F - c$ is a Keller map in $\mathbb{Z}_p[x]^n$ and $F^{-1}(c) = G^{-1}(0)$. If $b \in F^{-1}(c) = G^{-1}(0)$, then $G(b) = 0 \in (p\mathbb{Z}_p)^n$. So by Hensel's theorem b is completely determined by the element $\overline{b} \in (\mathbb{Z}_p/p\mathbb{Z}_p)^n$. Since there are at most p^n choices for $\overline{b}$ (for $\mathbb{Z}_p/p\mathbb{Z}_p \simeq \mathbb{F}_p$), there are also at most p^n choices for $b \in G^{-1}(0) = F^{-1}(c)$, which concludes the proof.   □

Theorem 4.5.8 *The Jacobian Conjecture is true if the unimodular conjecture is true for the p-adic integers, for almost all p.*

Proof

(i) It is well known that it suffices to prove the Jacobian Conjecture for Keller maps with integers coefficients ([117], Proposition 1.1.19). So let $F \in \mathbb{Z}[x]^n$ with $\det JF = 1$. We view F as a map from $\overline{\mathbb{Q}}^n$ to $\overline{\mathbb{Q}}^n$, where $\overline{\mathbb{Q}}$ is the algebraic closure of $\mathbb{Q}$. It suffices to show that this map is injective, because it then follows that F is invertible over $\overline{\mathbb{Q}}$ ([117], Theorem 4.2.1) and hence, since $\det JF = 1$, that F is invertible over $\mathbb{Z}$ ([117], Lemma 1.1.8).

(ii) Assume that $F(a) = F(b)$ with $a \neq b \in \overline{\mathbb{Q}}^n$. For infinitely many p, we can embed $\mathbb{Z}[a_1, \ldots, a_n, b_1, \ldots, b_n]$ into $\mathbb{Z}_p$ ([117], Theorem 10.3.1). Choose such a p for which the unimodular conjecture also holds for $\mathbb{Z}_p$ and consider $F : \mathbb{Z}_p^n \to \mathbb{Z}_p^n$. Since

$F(a) = F(b)$ with $a \neq b \in \mathbb{Z}_p^n$ and the unimodular conjecture holds for $\mathbb{Z}_p$, it follows from Theorem 4.5.5 that there exists a Keller map with coefficients in $\mathbb{Z}_p$ that has a fiber of at least $p^n + 1$ elements. This contradicts Theorem 4.5.7 and completes the proof.   □

Corollary 4.5.9 *The Jacobian Conjecture and the unimodular conjecture are equivalent.*

Proof   Follows immediately from Proposition 4.5.1 and Theorem 4.5.8.   □

Finally, we are able to give a new surprising formulation of the Jacobian Conjecture. Namely, let $F \in \mathbb{Z}_p[x]^n$ with $det\, JF = 1$. By reducing the coefficients of F mod $p\mathbb{Z}_p$, we obtain a polynomial map $\overline{F} : \mathbb{F}_p^n \to \mathbb{F}_p^n$. If the Jacobian Conjecture is true, it follows that $\overline{F}$ is a bijection, namely it follows from Proposition 1.1.12, [117], that the Jacobian Conjecture also holds for polynomial maps with coefficients in $\mathbb{Z}_p$. So F has an inverse G in $\mathbb{Z}_p[x]^n$. Reducing the equation $F \circ G = x$ mod $p\mathbb{Z}_p$, we obtain that $\overline{F}$ is a bijection with inverse $\overline{G}$. Now the next result shows that instead of showing that $\overline{F}$ is a bijection, it suffices to show that it is not the zero map!

Theorem 4.5.10 *The Jacobian Conjecture is equivalent to the following statement: for each $n \geq 1$, we have that for almost all prime numbers p each Keller map $F \in \mathbb{Z}_p[x]^n$ has the property that its induced map $\overline{F} : \mathbb{F}_p^n \to \mathbb{F}_p^n$ is not the zero map.*

Proof   Follows immediately from Proposition 4.5.1, Theorem 4.5.8, and Lemma 4.5.11 below.   □

Lemma 4.5.11 *$UC(\mathbb{Z}_p, n)$ is equivalent to: if $F \in \mathbb{Z}_p[x]^n$ is a Keller map, its induced map $\overline{F} : \mathbb{F}_p^n \to \mathbb{F}_p^n$ is not the zero map.*

Proof   Just observe that an element of $u \in \mathbb{Z}_p^n$ is unimodular if and only if $\overline{u} \in \mathbb{F}_p^n$ is unimodular (since $\mathbb{Z}_p$ is a local ring) or equivalently if $\overline{u} \neq 0$ in $\mathbb{F}_p^n$.   □

Consequently, in order to prove the Jacobian Conjecture, one "only" needs to show that the unimodular conjecture is true for local rings (since $\mathbb{Z}_p$ is local). To conclude this section, we will show that for local rings whose residue field is infinite the unimodular conjecture is true:

Proposition 4.5.12 *Let R be a local ring with maximal ideal m such that $k := R/m$ is infinite, then the unimodular conjecture holds for R. In particular, this is the case when R contains the rationals.*

Proof Let $F = (F_1, \ldots, F_n) \in R[x]^n$ be a Keller map. As in the proof of Lemma 4.5.11, it suffices to show that $\overline{F} : k^n \to k^n$ is not the zero map. However, if $\overline{F} : k^k \to k^n$ is the zero map, then the fact that k is infinite implies that $\overline{F_i} = 0$ for each i. So for all i all coefficients of F_i belong to the maximal ideal m, contradicting the hypothesis that $det\ JF = 1$. $\qquad\qquad\square$

4.6 A Mysterious Faulty Proof of the Jacobian Conjecture

Since I started my research related to the Jacobian Conjecture, I received hundreds of "proofs." Most of the time the error was easy to spot. However, a few years ago the situation was different.

On September 30, 2017, a paper was submitted to the *Journal of Pure and Applied Algebra* with the title *A proof of the Jacobian Conjecture*. The paper was sent to Wenhua Zhao, who was very busy at that time and therefore asked the advice of Pascal Adjamagbo and myself. We rather quickly spotted an error in the proof and reported it to Zhao who informed Chuck Weibel, the editor of *Journal of Pure and Applied Algebra*. A few days later, on October 6, the author wrote to Weibel:

> I express my thanks to the referee. By his comments, I could identify my wrong conclusion in the "proof" of the false statement and could improve my note. I would like to submit to you for publication in Journal of Pure and Applied Algebra the revised version.

Apart from its introduction, the revised version was about five pages that were very clearly written. On October 9, I started to look at the "proof" and could not detect an error. So I decided to ask the help of several friends. I rewrote the proof in my own way and split it up into three small parts that could be read independently. After a few days, parts 1 and 2 were checked and no error was found. The last part was still pending. In the meantime, Adjamagbo, who received the author's original proof, claimed that it could not be correct because the part where the author used the characteristic zero condition could be reproved without using this condition. Nevertheless, no one could point out where a possible error was located.

Inspired by Adjamagbo's observation, I succeeded in proving a lemma necessary in the crucial part of the proof, indeed without using the characteristic zero condition. This implied that the Jacobian Conjecture would be true in *any characteristic*, which is obviously false. On November 1, I sent out an email to several mathematicians containing the following lines:

> His proof, in which nobody can point out an error, leads to the fact that the Jacobian Conjecture is true. The proof of the lemma I added leads to the conclusion that his proof must be false. For me, this is one big mystery.

Two days later, I sent a message to David Wright, containing my one page proof of the crucial lemma that should contain the error and asked his help. The next day he replied by indicating more precisely where the error must be located. Then finally, on Sunday,

November 5, inspired by David's observation, I could detect the error. Immediately, I sent out an email to the author and everyone involved in the search, explaining the error. My email started as follows:

Dear Friends,

Thanks to the help of David Wright the error in the "proof" is finally found! The error is reflected by the following question: what have -1 and infinity in common?
Answer: they both can be written as an infinite sum, namely

$$1 + 2 + 2^2 + 2^3 + \cdots$$

In the 2-adic completion of $\mathbb{Q}$ it is -1 and in the usual completion infinity.
Now to the actual "proof" of JC:

After having discovered the error, I was able to give a very short "proof" of the Jacobian Conjecture based on the same error which was so mysteriously hidden in the revised paper of the author. Below you find my short "proof." It is now up to you to find the error. A hint is given in the email cited above. Have fun!

"Proof" of the Jacobian Conjecture

(i) We show $f : \mathbb{C}^n \to \mathbb{C}^n$ is injective. Let $f(a) = f(b)$, with $a, b \in \mathbb{C}^n$. Put $g(x) :=$ $f(x + a) - f(b)$. Then $g(0) = 0$ and $g(c) = 0$, where $c := b - a$. Also $det\, Jg(0) =$ $det\, Jf(a) \in \mathbb{C}^*$, since $det\, Jf \in \mathbb{C}^*$. Claim: $(x_1, \cdots, x_n) \subseteq (g_1, \cdots, g_n)$. So $g(c) =$ 0 implies $c = 0$, i.e., $b = a$.

(ii) To prove the claim let $A := \mathbb{C}[[y]] := \mathbb{C}[[y_1, \cdots, y_n]]$, $C := A[x]$, $F_j :=$ $g_j(x) - y_j$, $I := \sum_j CF_j$, and $S := \hat{C}$ the I-adic completion of C. Let $G :=$ $(F_1, \cdots, F_n, y_1, \cdots, y_n)$ in $A[[x]]^{2n}$. Since $det\, JG \in \mathbb{C}^*$ and $G(0, 0) = 0$, the formal inverse function theorem implies that there exists $h_j \in A[[x]]$ such that $h_j(F_1, \cdots, F_n) = x_j$ and $h_j(0, 0) = 0$ for all j. Let $a_j := h_j(x = 0) \in A$. Then $x_j - a_j \in \sum_{|\alpha| \geq 1} A F^\alpha \subseteq \sum SF_j$. Since $x_j - a_j \in \sum SF_j \cap C = I$ (the canonical map $C \to \hat{C} \to \hat{C}/\hat{I}$ has kernel I and $\hat{I} = I\hat{C} = IS$, since C is Noetherian). So $x_j - a_j \in \sum CF_j = \sum C(g_j(x) - y_j)$. Since $a_j(y = 0) = 0$ (for $h_j(0, 0) = 0$), we get $(x_1, \cdots, x_n) \subseteq \sum \mathbb{C}[x]g_j(x)$.

$\square$

Exercises for This Chapter

1. Let $E = A^n$ and $h : E \to E$ be an A-linear map. If $\omega \in \Lambda^n E$ is a volume form on E, show that $h^*\omega = (det\, h)\omega$.
2. Let ϕ be an R-endomorphism of the polynomial ring $A = R[x_1, \ldots, x_{2n}]$. Show that ϕ is an endomorphism of $P_n(R)$ if and only if

$$\{\phi(x_i), \phi(x_j)\} = \{x_i, x_j\}, \text{ for all } i, j.$$

3. Let $E := A^{2n}$ be the free A-module of rank $2n$. Put $v := e_1^* \wedge \cdots \wedge e_{2n}^*$ the standard volume form on E and let $\omega := \sum_{i=1}^{n} e_i^* \wedge e_{i+n}^*$. Show that $\omega^n = n!(-1)^{\frac{n(n-1)}{2}} v$.

4. Consider the Weyl algebra $A_1(\mathbb{Z}) = \mathbb{Z}[y_1, y_2]$. Show by induction on m that

$$[y_1^m, y_2^m] = \sum_{k=0}^{m-1} \frac{m!}{k!} \binom{m}{k} y_2^k y_1^k$$

for all $m \geq 1$.

5. Let k be a field of characteristic zero and ϕ a k-automorphism of $A_n(k)$ of degree $d \geq 1$. Show that ϕ^{-1} has degree $\leq d^{2n-1}$.

Notes

The proof of the equivalences of the Poisson, Dixmier, and Jacobian Conjectures is taken from the paper [2]. Crucial in this exposition is Proposition 4.3.8, which was inspired by the work of Belov-Kanel and Kontsevich in [8].

The Unimodular Conjecture was introduced in [120]. In the same paper, the p-adic formulations of the Jacobian Conjecture, Theorems 4.5.8, and 4.5.10 were obtained.

Mathieu–Zhao Spaces

5

5.1 Generalizing the Jacobian Conjecture

Throughout this section k denotes an *algebraically closed* field of characteristic zero. But, as the reader can check, most of the calculations done after Corollary 5.1.3 work equally well if k is just a commutative $\mathbb{Q}$-algebra.

In [117, Theorem 6.3.1] it is shown that it is sufficient to investigate the Jacobian Conjecture for polynomial maps of the form $F = x + H$, where each H_i is either zero or homogeneous of degree three. Furthermore, for these maps the Jacobian condition is equivalent to JH being nilpotent ([117, Lemma 6.2.11]). In June 2003 Michiel de Bondt and the author showed in [25] that one may additionally assume that JH is symmetric (see Theorem 5.1.1 below). By Poincaré's lemma [117, Lemma 1.3.53] a Jacobian matrix JH is symmetric if and only if there exists an $f \in k[x]$ such that $H = \nabla f = (f_{x_1}, \ldots, f_{x_n})^t$, i.e., H is a so-called **gradient map**.

Theorem 5.1.1 (de Bondt, van den Essen) *If the Jacobian Conjecture is true for all polynomial maps of the form $x + \nabla f : k^{2n} \to k^{2n}$, with $J(\nabla f)$ nilpotent and homogeneous, then the Jacobian Conjecture is true for all polynomial maps of the form $x + H : k^n \to k^n$ with JH nilpotent and homogeneous.*

The proof of this result is based on the following lemma. Recall that

$$J(\nabla f) = \left(\frac{\partial^2 f}{\partial x_i \partial x_j} \right) =: \mathscr{H}(f)$$

the **Hessian** of f.

© The Author(s), under exclusive license to Springer Nature Switzerland AG 2021
A. van den Essen et al., *Polynomial Automorphisms and the Jacobian Conjecture*,
Frontiers in Mathematics, https://doi.org/10.1007/978-3-030-60535-3_5

Lemma 5.1.2 *Let* $H = (H_1(x), \ldots, H_n(x)) \in k[x]^n$ *and let* $y_1, \ldots, y_n$ *be new variables. Let* $f = (-i) \sum_{j=1}^n y_j H_j(x + iy)$. *Then* JH *is nilpotent if and only if* $\mathscr{H}(f)$ *is nilpotent.*

Proof Let $S = (x - iy, y)$. Then $g := f \circ S = (-i) \sum_j y_j H_j(x)$ and

$$\nabla g = \left((-i) \sum_j y_j H_{j x_1}, \ldots, (-i) \sum_j y_j H_{j x_n}, (-i) H_1, \ldots, (-i) H_n \right).$$

So

$$\mathscr{H}(g) = J(\nabla(g)) = \begin{pmatrix} * & (-i) J H^t \\ (-i) J H & 0 \end{pmatrix}. \tag{5.1.1}$$

Also it follows from $g = f \circ S$ and Exercise 1 that

$$\mathscr{H}(g) = S_0^t \mathscr{H}(f)_{|S(x,y)} S_0 \tag{5.1.2}$$

where S_0 denotes the matrix corresponding to the linear map S. Now observe: $\mathscr{H}(f)$ is nilpotent if and only if $det \left(T I_{2n} - \mathscr{H}(f) \right) = T^{2n}$ if and only if $det S_0^t (T I_{2n} - \mathscr{H}(f)_{|S(x,y)}) S_0 = T^{2n}$ (since $det S_0 = 1$) if and only if $det (T S_0^t S_0 - \mathscr{H}(g)) = T^{2n}$ (by (5.1.2)). Since

$$S_0^t S_0 = \begin{pmatrix} I_n & -i I_n \\ -i I_n & 0 \end{pmatrix}$$

it follows from (5.1.1) that $\mathscr{H}(f)$ is nilpotent if and only if

$$det \begin{pmatrix} * & -i I_n T + i(JH)^t \\ -i I_n T + i J H & 0 \end{pmatrix} = T^{2n}.$$

Since the left-hand side equals $det (I_n T - (JH)^t) det (I_n T - JH)$, we see that $\mathscr{H}(f)$ is nilpotent if and only if $det (I_n T - JH) = T^n$ if and only if JH is nilpotent. $\square$

Proof of Theorem 5.1.1 Let $F = x + H$, where $H = (H_1(x), \ldots, H_n(x))$ with JH nilpotent and homogeneous. Let f be as in Lemma 5.1.2. Then $\mathscr{H}(f)$ is nilpotent. So by our hypothesis

$$G := (x_1 + f_{x_1}, \ldots, x_n + f_{x_n}, y_1 + f_{y_1}, \ldots, y_n + f_{y_n})$$

is invertible. Hence, with S as in the proof of Lemma 5.1.2, also $S^{-1} \circ G \circ S$ is invertible. An easy calculation shows that

$$S^{-1} \circ G \circ S = (F_1(x), \ldots, F_n(x), *, \ldots, *).$$

Let $(G_1(x, y), \ldots, G_n(x, y), *, \ldots, *)$ be its inverse. Then in particular

$$F_i(G_1(x, y), \ldots, G_n(x, y)) = x_i, \quad \text{for all } i.$$

Substituting $y = 0$ shows that F is invertible. □

Combining Theorem 5.1.1 with the classical Bass–Connell–Wright/Yagzhev reduction theorem ([117, Theorem 6.3.1]) we get

Corollary 5.1.3 *The following statements are equivalent:*

(i) *The Jacobian Conjecture.*
(ii) *The Jacobian Conjecture for polynomial maps of the form $x + \nabla f$, with $\mathcal{H}(f)$ nilpotent and f homogeneous of degree* 4.

So in order to study the Jacobian Conjecture one is naturally led to the question: what does the formal inverse of $x + \nabla f$ look like, in case f is homogeneous and $\mathcal{H}(f)$ is nilpotent? This question was answered by Wenhua Zhao in [135]. To describe his result we need several preparations. To avoid minus signs we write $F = x - H$ instead of $x + H$ from now on.

Recall that if R is a commutative ring and $g \in R[[x]]$ a formal power series in x over R, then the **order** of g, denoted $o(g)$, is the smallest degree of a monomial appearing in g if $g \neq 0$ and $o(g) = \infty$ if $g = 0$. More generally, if $H = (H_1, \ldots, H_n) \in R[[x]]^n$, then $o(H)$ denotes the minimum of the $o(H_i)$.

Let $H \in k[[x]]^n$ with $o(H) \geq 2$. Then the formal map $F = x - H$ satisfies $\det JF(0) = 1$, so F has a formal inverse. To study this inverse we introduce a new variable t. Let $A := k[t]$ and define

$$F_t := x - tH(x) \in A[[x]]^n.$$

Since $\det(J_x F_t)(0) = 1$ it follows from the formal inverse function theorem ([117, Theorem 1.1.2]) that F_t has a unique formal inverse, say G_t, in $A[[x]]^n$ of the form $x + M_t(x)$ with $o(M_t) \geq 2$. Setting $t = 0$ in the equation $F_t(G_t(x)) = x$ we get $G_0(x) = x$. So $M_t(x) = tN_t(x)$, for some $N_t(x) \in A[[x]]^n$. Then the equation $G_t(F_t(x)) = x$ implies that $x - tH(x) + tN_t(F_t(x)) = x$, whence

$$N_t(F_t(x)) = H(x) \tag{5.1.3}$$

and substituting $t = 0$ gives

$$N_0(x) = H(x).$$

By the chain rule we get $J N_t(F_t) \cdot J F_t = J H$. Since $J F_t = I - t J H$ this gives the following identity in the ring of formal power series in t over the matrix ring $M_n(k[[x]])$:

$$J N_t(F_t) = J H \cdot (I - t J H)^{-1} = \sum_{l=1}^{\infty} (J H)^l t^{l-1}. \qquad (5.1.4)$$

Differentiating (5.1.3) with respect to t and writing ∂_t instead of $\partial/\partial t$ we get $\partial_t(N_t)(F_t) - (J N_t)(F_t) \cdot H = 0$. Composing with G_t from the right and using (5.1.3) this gives

$$\partial_t(N_t) = J N_t \cdot N_t. \qquad (5.1.5)$$

From now on we assume that $J H$ is *symmetric*. So $H = \nabla f$, for some $f \in k[[x]]$, with $o(f) \geq 3$. It follows from (5.1.4) that $J N_t(F_t)$ is symmetric and hence so is $J N_t$. So there exists a unique $g_t \in A[[x]]$ with $o(g_t) \geq 3$ such that $N_t(x) = \nabla g_t$, whence

$$G_t = x + t \nabla g_t.$$

Since $N_0 = H$ we get $\nabla g_0 = \nabla f$, so

$$g_0 = f. \qquad (5.1.6)$$

From (5.1.5) we get $\nabla(\partial_t(g_t)) = J(N_t)\nabla g_t$. Also, writing $\langle, \rangle$ for the standard bilinear form, one easily verifies that

$$\nabla(\frac{1}{2} \langle \nabla g_t, \nabla g_t \rangle) = \mathcal{H}(g_t)\nabla g_t = J(N_t)\nabla g_t.$$

So $\nabla(\partial_t(g_t)) = \nabla(\frac{1}{2} \langle \nabla g_t, \nabla g_t \rangle)$, which implies that

$$\partial_t(g_t) = \frac{1}{2} \langle \nabla g_t, \nabla g_t \rangle. \qquad (5.1.7)$$

Now we address the question: what does the formal inverse of $x - \nabla f$ look like, in case $\mathcal{H}(f)$ is nilpotent? Therefore we need one more lemma:

Lemma 5.1.4 *Let $f \in k[[x]]$ with $o(f) \geq 3$. Then $\mathcal{H}(f)$ is nilpotent if and only if $\Delta g_t = 0$, where $\Delta := \sum_i \partial_i^2$ denotes the Laplace operator.*

Proof   Since $JN_t = J(\nabla g_t) = \mathcal{H}(g_t)$ we get $Tr\,JN_t = Tr\,\mathcal{H}(g_t) = \Delta g_t$. It then follows from (5.1.4) by taking traces that

$$(\Delta g_t)(F_t) = \sum_{l=1}^{\infty} Tr\,\mathcal{H}(f)^l t^{l-1} \tag{5.1.8}$$

which gives the desired result, since an $n \times n$ matrix over a domain is nilpotent if and only if the traces of its first n powers are zero (or use Example 5.2.4 in case k is a commutative $\mathbb{Q}$-algebra).   □

Theorem 5.1.5 (Zhao) *Let $f \in k[[x]]$ with $o(f) \geq 3$ and $\mathcal{H}(f)$ nilpotent. Then*

$$g_t^l = l! \sum_{m=0}^{\infty} \frac{t^m}{2^m m!(m+l)!} \Delta^m(f^{m+l}) \text{ for all } l \geq 1. \tag{5.1.9}$$

Proof   Let $U := \exp(sg_t)$, where s is a new variable. Then $U(t = 0) = \exp(sf)$ and, using (5.1.7), we get

$$\partial_t(U) = \frac{s}{2}\langle \nabla g_t, \nabla g_t \rangle U. \tag{5.1.10}$$

Also one verifies that

$$\Delta U = s \sum_i \partial_i(\partial_i(g_t)U) = s\Delta(g_t)U + s^2 \sum_i \partial_i(g_t)^2 U. \tag{5.1.11}$$

Since $\mathcal{H}(f)$ is nilpotent, we get $\Delta g_t = 0$ by Lemma 5.1.4. So (5.1.11) gives

$$\Delta U = s^2 \sum_i \partial_i(g_t)^2 U = s^2 \langle \nabla g_t, \nabla g_t \rangle U. \tag{5.1.12}$$

Combining (5.1.10) and (5.1.12) and the equality $U(t = 0) = \exp(sf)$ we obtain

$$\partial_t(U) = \frac{1}{2s} \Delta U \text{ and } U(t = 0) = \exp(sf). \tag{5.1.13}$$

So U is the unique power series in t which is solution of (5.1.13). But as one readily verifies also $V = \exp(\frac{t}{2s}\Delta)(\exp(sf))$ satisfies the conditions in (5.1.13). So $U = V$. Finally comparing the coefficients of s^l in both U and V for each $l \geq 1$ we get (5.1.9).   □

Corollary 5.1.6 *Let $f \in k[[x]]$ with $o(f) \geq 3$ and $\mathcal{H}(f)$ nilpotent. Then $\Delta^m(f^m) = 0$ for all $m \geq 1$.*

Proof Apply Δ to (5.1.9) (with $l = 1$) and use Lemma 5.1.4. $\square$

Now we will show that the converse also holds.

Proposition 5.1.7 *Let $r \geq 1$. If $\Delta^m f^m = 0$ for all $1 \leq m \leq r$, then $Tr\, \mathscr{H}(f)^m = 0$, for all $1 \leq m \leq r$.*

Proof By induction on r. The case $r = 1$ is obvious since $Tr\, \mathscr{H}(f) = \Delta f$. So let $r \geq 2$. Assume $\Delta^m f^m = 0$ for all $1 \leq m \leq r + 1$. Then the induction hypothesis implies that $Tr\, \mathscr{H}(f)^m = 0$ for all $1 \leq m \leq r$. So by (5.1.8) we have $\Delta g_t \equiv Tr\, \mathscr{H}(f)^{r+1} t^r (mod\ t^{r+1})$. Consequently we get

$$Tr\, \mathscr{H}(f)^{r+1} \equiv \frac{1}{r!} \partial_t^r (\Delta g_t)(mod\ t).$$

Now we claim that $\partial_t^r (\Delta g_t) \equiv c \Delta^{r+1} g_t^{r+1} (mod\ t)$ for some $c \in k^*$. Assuming this claim for a moment, it follows from (5.1.6) that

$$Tr\, \mathscr{H}(f)^{r+1} = \frac{c}{r!} \Delta^{r+1} f^{r+1} = 0.$$

Finally, the claim follows by taking $l = r$ in the next lemma and applying Δ. In that lemma we use the symbol $*$ to denote a unit in k. $\square$

Lemma 5.1.8 *Let $r \geq 1$ and $Tr\, \mathscr{H}(f)^i = 0$ for all $1 \leq i \leq r$. Then*

$$\partial_t^l g_t \equiv * \Delta^l g_t^{l+1} \ (mod\ t^{(r+1)-l}) \ \text{for all } 1 \leq l \leq r. \tag{5.1.14}$$

Proof By induction on l. The case $l = 1$: by the hypothesis and (5.1.8) we have $\Delta g_t \equiv 0 (mod\ t^r)$. So, by looking at the coefficient of s^2 in (5.1.11), we get

$$\Delta g_t^2 \equiv * \langle \nabla g_t, \nabla g_t \rangle (mod\ t^r).$$

Consequently, using (5.1.7), we obtain

$$\partial_t g_t = \frac{1}{2} \langle \nabla g_t, \nabla g_t \rangle \equiv * \Delta g_t^2 (mod\ t^r) \tag{5.1.15}$$

which proves the case $l = 1$. So assume the case $1 \leq l \leq r - 1$ is proved. We prove the case $l + 1$. Therefore, applying ∂_t to (5.1.14), gives

$$\partial_t^{l+1} g_t \equiv * \Delta^l \partial_t g_t^{l+1} (mod\ t^{(r+1)-(l+1)}). \tag{5.1.16}$$

Now observe that by using (5.1.15) we obtain

$$\partial_t g_t^{l+1} = (l+1)g_t^l \partial_t g_t \equiv *\Delta g_t^{l+2} (mod \ t^r).$$

Substituting this in (5.1.16) gives the case $l+1$. □

Theorem 5.1.9 (Zhao) *The Jacobian Conjecture is equivalent to each of the following statements.*

(i) *If f is homogeneous of degree ≥ 3 such that $\Delta^m f^m = 0$ for all $m \geq 1$, then $\Delta^m f^{m+1} = 0$ for all large m.*

(ii) *If f is homogeneous of degree 4 such that $\Delta^m f^m = 0$ for all $m \geq 1$, then $\Delta^m f^{m+1} = 0$ for all large m.*

Proof Let f be homogeneous of degree 4. If $\mathscr{H}(f)$ is nilpotent, or equivalently (using Corollary 5.1.6 and Proposition 5.1.7), if $\Delta^m f^m = 0$ for all positive integers m, it follows from (5.1.9) with $l = 1$, by substituting $t = 1$, that the formal inverse of $x - \nabla f$ is of the form $x + \nabla g$, where

$$g = \sum_{m=0}^{\infty} \frac{1}{2^m m!(m+1)!} \Delta^m (f^{m+1}).$$

Then the result follows from Corollary 5.1.3. □

The reader may wonder what is so special about the Laplace operator? The answer is, that it is not so special after all. In fact one can show that if the statement in Theorem 5.1.9 holds, then it holds with Δ replaced by *any* quadratic homogeneous differential operator with constant coefficients in $\partial_1, \ldots, \partial_n$ (Exercise 5).

Furthermore, it was shown in [124] that statement i) of Theorem 5.1.9 is equivalent to the following conjecture (see also Theorem 5.6.2):

Vanishing Conjecture *If $f \in k[x]$ is homogeneous and such that $\Delta^m f^m = 0$, for all $m \geq 1$, then for every $g \in k[x]$ also $\Delta^m (gf^m) = 0$ for all large m.*

Both observations above led Zhao to make the following more general conjecture, where also the homogeneity condition on f is dropped.

Generalized Vanishing Conjecture (GVC(n)) *Let Λ be any differential operator with constant coefficients, i.e., $\Lambda \in k[\partial_1, \ldots, \partial_n]$. If $f \in k[x]$ is such that $\Lambda^m f^m = 0$ for all $m \geq 1$, then for each $g \in k[x]$ also $\Lambda^m (gf^m) = 0$ for all large m.*

Since for $\Lambda = \Delta$ we obtain the Vanishing Conjecture, which as observed above is equivalent to the Jacobian Conjecture, one can view the Generalized Vanishing Conjecture as a *family of "new Jacobian Conjectures,"* one for each $\Lambda \in k[\partial_1, \ldots, \partial_n]$.

The main obstruction in studying the Generalized Vanishing Conjecture is the fact that the operators Λ and f do not commute. This led Zhao to the following even more general conjecture. The reader is referred to [118] for an account of its history. To describe this conjecture we introduce n new commuting variables $\zeta_1, \ldots, \zeta_n$ and consider the $2n$-variable polynomial ring $k[x, \zeta] := k[x_1, \ldots, x_n, \zeta_1, \ldots, \zeta_n]$. Furthermore let D be the following set of commuting differential operators on $k[x, \zeta]$:

$$\{\partial_{x_1} - \zeta_1, \ldots, \partial_{x_n} - \zeta_n\}.$$

Finally put

$$Im\, D := \sum_i (\partial_{x_i} - \zeta_i)k[x, \zeta].$$

Special Image Conjecture (SIC(n)) *Let k be a commutative ring contained in a $\mathbb{Q}$-algebra. If $f \in k[x, \zeta]$ is such that $f^m \in Im\, D$ for all $m \geq 1$, then for every $g \in k[x, \zeta]$: $gf^m \in Im\, D$ for all large m.*

Theorem 5.1.10 (Zhao) *SIC(n) implies $GVC(n)$.*

To prove this theorem we introduce the k-linear map $\mathscr{E} : k[x, \zeta] \to k[x]$, defined by $\mathscr{E}(\zeta^a x^b) = \partial_x^a(x^b)$.

Proof of Theorem 5.1.10 Let $\Lambda = \Lambda(\partial)$ be a differential operator with constant coefficients and $f \in k[x]$ such that $\Lambda^m(f^m) = 0$ for all positive m. Let $g \in k[x]$. We must show that $\Lambda^m(gf^m) = 0$ for all large m. Therefore put $f(x, \zeta) = \Lambda(\zeta)f(x)$. It then follows that $\mathscr{E}(f(x, \zeta)^m) = \mathscr{E}(\Lambda(\zeta)^m f(x)^m) = \Lambda(\partial)^m f(x)^m = 0$ for all positive m. Since by Proposition 5.1.11 below $ker\, \mathscr{E} = Im\, D$ we deduce that for all positive m $f(x, \zeta)^m \in Im\, D$. Since we assume $SIC(n)$ we conclude that $g(x)f(x, \zeta)^m \in Im\, D = ker\, \mathscr{E}$ for all large m. So $\mathscr{E}(\Lambda(\zeta)^m g(x) f(x)^m) = 0$ for all large m, i.e., $\Lambda^m(gf^m) = 0$ for all large m. So $GVC(n)$ holds. $\qquad\square$

Proposition 5.1.11 *$Im\, D = ker\, \mathscr{E}$.*

To prove this result we use the following well-known result from the theory of $\mathfrak{D}$-modules:

Proposition 5.1.12 *Let M be an $A_n(k) = k[t_1, \ldots, t_n, \partial_1, \ldots, \partial_n]$-module such that each ∂_i is locally nilpotent on M. Then each m in M can be written uniquely in the form $m =$*

$\sum t^a m_a$, where each m_a belongs to $N = \bigcap_i ker(\partial_i, M)$. In particular $m \in \sum t_i M$ if and only if $m_0 = 0$.

Proof of Proposition 5.1.11 Apply Proposition 5.1.12 to $M = k[x, \zeta]$, which is an $A_n(k)$-module by defining

$$t_i f(x, \zeta) := (\zeta_i - \partial_{x_i}) f(x, \zeta) \text{ and } \partial_i f(x, \zeta) := \partial_{\zeta_i} f(x, \zeta).$$

Then $N = \bigcap_i ker(\partial_{\zeta_i}, M) = k[x]$. So $f(x, \zeta) \in k[x, \zeta]$ can be written uniquely in the form

$$f(x, \zeta) = \sum (\zeta_i - \partial_{x_i})^a f_a(x) \tag{5.1.17}$$

for some $f_a(x) \in k[x]$, and in particular $f(x, \zeta) \in \sum (\partial_{x_i} - \zeta_i) k[x, \zeta] = Im\, D$ if and only if $f_0(x) = 0$. Finally observe that (5.1.17) implies that $\mathscr{E}(f(x, \zeta)) = f_0(x)$. So we get $f(x, \zeta)$ belongs to $Im\, D$ if and only if $f(x, \zeta)$ belongs to $ker\, \mathscr{E}$, which completes the proof. $\square$

The proof of Proposition 5.1.12 follows by induction on n, using the following lemma (see Exercise 6).

Lemma 5.1.13 *If M is an $A_1(k) = k[t, \partial]$-module and $m \in M$ such that $\partial^r m = 0$ for some $r \geq 1$, then $m = m_0 + t m_1 + \cdots + t^{r-1} m_{r-1}$ for some $m_i \in N := ker(\partial, M)$, which are uniquely determined.*

Proof The uniqueness follows easily by applying ∂ sufficiently many times. So let $\partial^r m = 0$. We use induction on r. The case $r = 1$ is obvious, so let $r \geq 2$. Since $\partial^{r-1}(\partial m) = 0$, the induction hypothesis implies that

$$\partial m = m_0 + t m_1 + \cdots + t^{r-2} m_{r-2}$$

for some $m_i \in N$. Now let

$$g := \int \partial m \, dt = t m_0 + \frac{1}{2} t^2 m_1 + \cdots + \frac{1}{r-1} t^{r-1} m_{r-2}.$$

Then $\partial g = \partial m$, so $m - g \in ker\, \partial = N$. Say $m - g = m'_0$ for some $m'_0 \in N$. Now the result follows using the definition of g. $\square$

Exercises for Sect. 5.1

1. Let $f \in k[x]$ and $A \in \mathrm{GL}_n(k)$. Define $g := f(Ax)$. Show that

$$\mathscr{H}(g) = A^t \mathscr{H}(f)_{|Ax} A.$$

2. Let $f \in k[x]$ of degree $d \geq 1$ and t be a new variable. Define

$$f^* := t^d f(\frac{x}{t}) \in k[x, t].$$

   (i) Let $e \geq 1$ and $\Lambda \in k[\partial]_{(e)}$, the set of differential operators homogeneous of degree e. Show that

$$\Lambda f^* = t^{d-e}(\Lambda f)(\frac{x}{t}).$$

   (ii) Let $m, N \in \mathbb{N}$. Deduce from i) that $\Lambda^m f^N = 0$ if and only if $\Lambda^m (f^*)^N = 0$.
3. Let k be a field of characteristic zero, $1 \leq r \leq n$ and $\Delta_r := \partial_1^2 + \cdots + \partial_r^2$. Let $VC(k, r)$ denote the following statement: if $f \in k[x_1, \ldots, x_n]$ is such that $\Delta_r^m f^m = 0$ for all $m \geq 1$, then $\Delta_r^m f^{m+1} = 0$ for all large m.
   Show that $VC(\mathbb{C}, n)$ implies $VC(k, r)$ [Use Lefschetz' principle].
4. Let $\Lambda \in k[\partial]_{(2)}$.
   (i) Show that there exists a unique symmetric matrix $A \in \mathrm{M}_n(k)$ such that $\Lambda = \partial^t A \partial$, where $\partial^t = (\partial_1, \ldots, \partial_n)$.
   (ii) Assume $k = \mathbb{C}$. For $U \in \mathrm{GL}_n(\mathbb{C})$ put $x' = U^t x$. Show that $\partial' = U^{-1}\partial$ and deduce from i) that there exists $U \in \mathrm{GL}_n(\mathbb{C})$ such that $\Lambda = \partial_1'^2 + \cdots + \partial_r'^2$, where $r = rank\ A$.
5. Let $\Lambda \in \mathbb{C}[\partial]_{(2)}$ and let $HVC(\Lambda, n)$ denote the following statement: if $f \in \mathbb{C}[x]$ is homogeneous such that $\Lambda^m f^m = 0$ for all $m \geq 1$, then $\Lambda^m f^{m+1} = 0$ for all large m.
   Show that $HVC(\Delta, n)$ implies $HVC(\Lambda, n)$ [Use Exercise 4 to reduce to the case that $\Lambda = \partial_1^2 + \cdots + \partial_r^2$, with $1 \leq r \leq n$. If $r \leq n - 1$ use Exercises 2 and 3].
6. Prove Proposition 5.1.12 by induction on n [Observe that $N_n := ker\ \partial_n$ is a left A_{n-1}-module and use Lemma 5.1.13].

Notes

Most results in this section were obtained by Wenhua Zhao in [135]. Symmetric Jacobian matrices in connection with the Jacobian Conjecture were first studied by Washburn and the author in [123]. It was this paper which formed the starting point that led Michiel de Bondt and the author to the symmetric reduction theorem (Theorem 5.1.1). This result was presented by the author at the Special Session of Affine Algebraic Geometry at the first joint AMS-RSME meeting, Seville, Spain, June 18–21, 2003. A similar result was independently obtained by G. Meng in [84]. It was first submitted to arXiv on August 28, 2003 and revised on January 31, 2005. In his thesis [20], de Bondt strongly improved upon

Theorem 5.1.1 by showing that one may even assume that the Jacobian matrix is symmetric with respect to the diagonal, the antidiagonal, and both the vertical and horizontal axes going through the center of the matrix. For the precise result we refer to [22, Theorem 2.5]. In the paper [26] it was shown that statement i) of Theorem 5.1.9 holds if $n \leq 5$. In other words that the Jacobian Conjecture holds for all polynomial maps of the form $x + \nabla f$, in case f is homogeneous of degree ≥ 3 and $n \leq 5$. The proof of this result is based on the paper [24].

5.2   Mathieu–Zhao Spaces: Definition and Examples

Throughout this section R denotes a commutative ring and A an associative, but not necessarily commutative, algebra over R. We always assume that A contains 1.

Let V be any subset of A. Then we define the **radical of** V, denoted $r(V)$, as the set of a in A such that $a^m \in V$ for **all large** m, i.e., there exists $N \geq 1$ such that $a^m \in V$ for all $m \geq N$ (N may depend on a). As we will see below, the radical can have a very wild structure. For example if V is an R-submodule of A, the radical of V need not even be additive (see Example 5.2.4).

We say that an element a in A has the **left (respectively right) Mathieu–Zhao property with respect to** V if for every b in A: $ba^m \in V$ (respectively $a^m b \in V$) for all large m. If for every b, c in A we have that $ba^m c \in V$ for all large m, we say that a has the **Mathieu–Zhao property with respect to** V. To simplify this presentation we will only give the definitions for the "left" case and will write MZ instead of Mathieu–Zhao from now on.

The set of all a in A which have the left MZ-property with respect to V is called the **(left) strong radical of** V, denoted $lsr(V)$. Most of the time, when no confusion is possible, for example when A is commutative, we will simply write $sr(V)$ and call it the **strong radical of** V.

Observe that if $a \in lsr(V)$, then in particular $1 \cdot a^m \in V$ for all large m, i.e., $a \in r(V)$. So $lsr(V) \subseteq r(V)$. In case equality holds such a space will be called a left Mathieu–Zhao space. More precisely:

Definition 5.2.1 Let M be an R-submodule of A. Then M is called a left Mathieu–Zhao space of A (MZ-space) if $lsr(M) = r(M)$, in other words if all elements of the radical of M have the left MZ-property with respect to M.

Obviously, if M is a left MZ-space, then all a in A such that $a^m \in M$ for all $m \geq 1$ have the left MZ-property with respect to M. However, the converse also holds:

Proposition 5.2.2 *If all $a \in A$ such that $a^m \in M$ for all $m \geq 1$ have the left MZ-property with respect to M, then all elements of $r(M)$ have the left MZ-property with respect to M, i.e., M is a left MZ-space.*

Proof Let $a \in r(M)$. So there exists $N \geq 1$ such that $a^m \in M$ for all $m \geq N$. Then in particular $(a^N)^m \in M$ for all $m \geq 1$. Let $b \in A$. Then the hypothesis implies that for each $0 \leq i \leq N - 1$ there exists an N_i such that $(ba^i)(a^N)^m \in M$ for all $m \geq N_i$. Hence, if N' is the maximum of all N_i, then $ba^{Nm+i} \in M$ for all $m \geq N'$ and all $0 \leq i \leq N - 1$. Since by Euclidean division each $n \geq NN'$ can be written in the form $Nm + i$, with $m \geq N'$ and $0 \leq i \leq N - 1$, it follows that $ba^n \in M$ for all $n \geq NN'$. $\square$

The notion of a Mathieu subspace was introduced by Zhao in [138]. His definition was motivated by the following conjecture of Olivier Mathieu in [82]. The name was later changed in [119] to Mathieu–Zhao space in order to honor Zhao for his great contributions.

Mathieu Conjecture *Let G be a compact connected real Lie group with Haar measure σ. Let f be a complex valued G-finite function on G such that $\int_G f^m \, d\sigma = 0$ for all $m \geq 1$. Then for every G-finite function g on G also $\int_G g f^m \, d\sigma = 0$ for all large m.*

Here a function f is called G-**finite** if the $\mathbb{C}$-vector space generated by the elements of the orbit $G \cdot f$ is finite dimensional.

Using Proposition 5.2.2 the Mathieu Conjecture can be reformulated as follows: let A be the $\mathbb{C}$-algebra of complex G-finite functions on G. Then the subspace of f's such that $\int_G f \, d\sigma = 0$ is an MZ-space of A.

The importance of Mathieu's conjecture comes from the fact that it implies the Jacobian Conjecture, as was shown in [82]. Since its formulation only one nontrivial case has been solved, namely the case that G is abelian: this result is the so-called **Duistermaat–van der Kallen theorem** which will be discussed in Sect. 5.5.

Also the Special Image Conjecture can be reformulated in terms of MZ-spaces, namely one easily verifies that this conjecture is equivalent to the statement that $Im\ D$ is an MZ-space of $k[x, \zeta]$. We return to this conjecture in Sect. 5.7.

Before we start studying MZ-spaces in more detail we discuss several examples. First of all, every left ideal I of A is a left MZ-space of A: for if $a^m \in I$, for all $m \geq N$, then $a^N \in I$, so for every $b \in A$ also $ba^m = (ba^{m-N})a^N \in I$ for all $m \geq N$, by the left ideal property. Consequently the concept of a left MZ-space is a *generalization* of the notion of a left ideal.

To verify if a given submodule M of A is a left MZ-space one first needs to compute the radical of M and then check if all its elements have the left MZ-property. In many cases the elements of the radical consist of nilpotent elements only. This gives a first class of examples of left (right and two sided) MZ-spaces:

Example 5.2.3 Let M be an R-submodule of A such that $r(M) \subseteq \mathfrak{n}(A)$, where $\mathfrak{n}(A)$ is the set of nilpotent elements of A. Then M is a left MZ-space of A.

Proof Let $a \in r(M)$. Then by the hypothesis $a^N = 0$ for some $N \geq 1$. Hence $ba^m = (ba^{m-N})a^N = 0 \in M$ for all $m \geq N$. So M is a left MZ-space of A. $\square$

Example 5.2.4 Let R be contained in a $\mathbb{Q}$-algebra and $A := M_n(R)$, the set of $n \times n$ matrices over R. Let $M := \{a \in A \mid Tr\, a \in \mathfrak{n}(R)\}$. Then $r(M) = \mathfrak{n}(A)$. So M is a left MZ-space of A.

Proof If $a \in \mathfrak{n}(A)$, then $a^N = 0$ for some $N \geq 1$. So $Tr\, a^m = 0$ for all $m \geq N$. Hence $a \in r(M)$. So it suffices to show that $r(M) \subseteq \mathfrak{n}(A)$. Then the result follows from Example 5.2.3. So let $a \in r(M)$. Since R is contained in a $\mathbb{Q}$-algebra S, we can view a inside $M_n(S)$ and may assume that R is a $\mathbb{Q}$-algebra. Furthermore replacing R by the $\mathbb{Q}$-algebra generated by the matrix elements a_{ij} of a, we may also assume that R is Noetherian.

To show the inclusion above we first assume that R is a domain. Since R is a $\mathbb{Q}$-algebra its characteristic is zero. Then it is well known that if for a matrix $a \in M_n(R)$ the traces $Tr\, a^m$ are zero for all $m \geq N$ and some N, then a is nilpotent.

Now the general case: since R is Noetherian $\mathfrak{n}(R)$ is a finite intersection of prime ideals $\mathfrak{p}_i$. For each i consider the domain $R_i := R/\mathfrak{p}_i$, which is also a $\mathbb{Q}$-algebra. It then follows, reducing modulo $\mathfrak{p}_i$, that $\bar{a}^{N_i} = 0$ in $M_n(R_i)$, for some $N_i \geq 1$. So if N is the maximum of the N_i, then all matrix elements of a^N belong to the intersection of all $\mathfrak{p}_i$, i.e., all matrix elements of a are nilpotent. Consequently, since R is Noetherian, a is nilpotent.   □

Example 5.2.5 Let R be contained in a $\mathbb{Q}$-algebra and G a finite group. Let $A := R[G]$ be the group ring of G over R. So every element of $R[G]$ is a finite sum of the form $\sum_{g \in G} f_g g$, with $f_g \in R$. Let e denote the neutral element of G and set

$$M := \{f \in R[G] \mid f_e \in \mathfrak{n}(R)\}.$$

Then $r(M) = \mathfrak{n}(A)$. So M is a left MZ-space of A.

Proof

(i) The elements $g \in G$ form an R-basis of the free R-module $R[G]$. If $f = \sum_g f_g g \in R[G]$, then the left multiplication map $\lambda(f) : R[G] \to R[G]$, defined by $\lambda(f)(h) = fh$ for all $h \in R[G]$, is R-linear. Using the R-basis above one verifies that $Tr\, \lambda(g) = 0$ for all $g \in G$, $g \neq e$. So

$$Tr\, \lambda(f) = \sum_g f_g Tr\, \lambda(g) = f_e Tr\, \lambda(e) = nf_e,$$

where $n = \#G$.

(ii) Now let $f \in r(M)$. Then $(f^m)_e \in \mathfrak{n}(R)$ for all large m. So by i) (applied to f^m) we get $Tr\, \lambda(f^m) \in \mathfrak{n}(R)$ for all large m. Since $\lambda(f^m) = \lambda(f)^m$ we obtain that $Tr\, \lambda(f)^m \in \mathfrak{n}(R)$ for all large m. Hence, as shown in Example 5.2.4, $\lambda(f)$ is nilpotent. So $\lambda(f)^N = 0$ for some $N \geq 1$. Consequently $\lambda(f^N) = 0$, which implies that

$f^N = 0$, since λ is injective. So $r(M) \subseteq \mathfrak{n}(A)$. The converse inclusion is obvious. Finally, the last statement follows from Example 5.2.3.   □

As remarked before, in various cases $r(M)$ is contained in the nilradical of the ring in question. However, if this is the case, the proof is often difficult, even if $r(M) = 0$. The next two examples illustrate this fact. For the details we refer to Sect. 5.4, but the reader is invited to find new proofs!

Example 5.2.6 Let $A = \mathbb{C}[t]$, the univariate polynomial ring over the complex numbers and

$$M = \{f(t) \in \mathbb{C}[t] \mid \int_0^1 f(t)\, dt = 0\}.$$

Then $r(M) = 0$, i.e., if $f(t) \in \mathbb{C}[t]$ is such that $\int_0^1 f(t)^m\, dt = 0$ for all large m, then $f(t) = 0$.

Example 5.2.7 Let $A = \mathbb{C}[t]$ and $\mathcal{L} : \mathbb{C}[t] \to \mathbb{C}$ defined by

$$\mathcal{L}(f(t)) = \int_0^\infty e^{-t} f(t)\, dt.$$

Then $r(\ker \mathcal{L}) = 0$.

From the definition of the gamma function it follows that $\mathcal{L}(t^n) = n!$ Therefore Example 5.2.7 can be generalized as follows:

Factorial Conjecture *Let R be contained in a $\mathbb{Q}$-algebra and $\mathcal{L} : R[x] \to R$ be the R-linear map defined by*

$$\mathcal{L}(x_1^{i_1} \cdots x_n^{i_n}) = i_1! \cdots i_n!$$

Then $r(\ker \mathcal{L}) \subseteq \mathfrak{n}(R)$.

Using the same techniques as used in the proof of Example 5.2.4 the Factorial Conjecture can be reduced to the case $R = \mathbb{C}$ (Exercise 1). However, even with this reduction this conjecture is open for all $n \geq 2$. Only for some special polynomials the statement of the conjecture has been proved (see for example [127]). On the other hand, ample experiments with the computer algebra system MAPLE suggest that the following, even stronger conjecture, might be true:

Strong Factorial Conjecture *Let $f \in \mathbb{C}[x]$ contain at most $N \geq 1$ nonzero monomials. If there exists an $m \geq 0$ such that the N consecutive powers $f^m, f^{m+1}, \ldots, f^{m+N-1}$ belong to $\ker \mathcal{L}$, then $f = 0$.*

Not much is known about this conjecture, even the case of one variable is still open. A first study of this conjecture appeared in the thesis of Brady Rocks ([105]). One of the results he proves is the following (Theorem 3.3): if $f = l^r$ for some $r \geq 1$ and $l = \lambda_1 x_1 + \cdots + \lambda_n x_n$ with $\lambda_i \in \mathbb{C}$, then f satisfies the statement of the strong factorial conjecture. On the other hand for f's of the form

$$f = x_1 \cdots x_n \cdot (\lambda_1 x_1 + \cdots + \lambda_n x_n) \tag{5.2.1}$$

we do not know if they satisfy the strong factorial conjecture. In fact, as is shown in [42], this question is related to the following conjecture of Jean-Philippe Furter ([51]):

Furter's Rigidity Conjecture (R(n)) Let $a(t) \in \mathbb{C}[t]$ of degree $\leq n + 1$ be such that $a(t) \equiv t \pmod{t^2}$. If n consecutive terms of the formal inverse of $a(t)$ vanish, then $a(t) = t$.

To explain the relation between this conjecture and the strong factorial conjecture, observe that a polynomial as described in Furter's conjecture can be written in the form $a(t) = t(1 - \lambda_1 t) \cdots (1 - \lambda_n t)$, for some $\lambda_i \in \mathbb{C}$. It is proven in [51] that the formal inverse $b(t)$ of $a(t)$ is given by

$$b(t) = t(1 + \sum_{m \geq 1} \frac{u_m}{m + 1} t^m) \in \mathbb{C}[[t]],$$

where for $m \geq 1$ the complex numbers u_m are given by

$$u_m = \sum_{j_1 + \cdots + j_n = m} \binom{m + j_1}{m} \cdots \binom{m + j_n}{m} \lambda_1^{j_1} \cdots \lambda_n^{j_n}.$$

Now consider the polynomial f defined in (5.2.1). Then it is left to the reader to verify that $\mathfrak{L}(f^m) = (m!)^{n+1} u_m$ and to deduce that $R(k)$ holds for all $1 \leq k \leq n$ if and only if all f's of the form (5.2.1) satisfy the strong factorial conjecture (See Exercise 2).

In the first section we introduced the Special Image Conjecture and showed how it implies the Jacobian Conjecture. The next two examples will shine a new light on the Jacobian condition and the invertibility of a homogeneous polynomial map.

Let R be a commutative $\mathbb{Q}$-algebra and $\mathscr{E} : R[x, \zeta] \to R[x]$ the R-linear map defined by $\mathscr{E}(\zeta^a x^b) = \partial^a(x^b)$. Let $H = (H_1(x), \ldots, H_n(x)) \in R[x]^n$ be homogeneous of degree $d \geq 2$ and $F := x - H$. Put

$$f(x, \zeta) := \zeta_1 H_1(x) + \cdots + \zeta_n H_n(x).$$

Example 5.2.8 $f(x, \zeta) \in r(\ker \mathscr{E})$ if and only if JH is nilpotent if and only if $\det JF \in R[x]^*$.

Proof Since the last equivalence is well known we only prove the first one. We use the classical **Abhyankar–Gurjar inversion formula**: let G be the formal inverse of F and $j(F) = det\ JF$. Then for any formal power series $u(x) \in R[[x]]$ we have

$$\sum_{m\geq 0}\sum_{|a|=m} \frac{1}{a!}\partial^a\big(H^a(x)j(F)u(x)\big) = u(G(x)).$$

Since $F(G(x)) = x$ the chain rule gives $j(F)(G)j(G) = 1$, so $j(F)^{-1}(G) = j(G)$. Hence, taking $u(x) = j(F)^{-1}$ in the formula above, we get

$$\sum_{m\geq 0}\sum_{|a|=m} \frac{1}{a!}\partial^a(H^a(x)) = j(G).$$

Now observe that the multinomial formula implies that

$$\mathscr{E}(f(x,\zeta)^m) = \mathscr{E}\left(\sum_{|a|=m} \frac{m!}{a!}\zeta^a H^a(x)\right) = m!\sum_{|a|=m} \frac{1}{a!}\partial^a(H^a(x)).$$

Since each component $\partial^a(H^a(x))$ is homogeneous of degree $(d-1)|a|$, it follows from the last two formulas that $j(G) \in R[x]$ if and only if $\mathscr{E}(f(x,\zeta)^m) = 0$ for all large m. So in order to prove the desired result we need to show that $j(G) \in R[x]$ if and only if JH is nilpotent, or equivalently that $j(F) \in R[x]^*$.

First, if $j(G) \in R[x]$, then $j(G)(F) \in R[x]$. So $j(G)(F)j(F) = 1$ implies that $j(F) \in R[x]^*$. Conversely, assume that $j(F) \in R[x]^*$. Let $\mathfrak{n}$ be the nilradical of R, $\overline{R} = R/\mathfrak{n}$ and $\overline{F}, \overline{G}$ be obtained by reducing F and G modulo $\mathfrak{n}$. Then $j(\overline{F}) \in \overline{R}^*$. So from $j(\overline{F})(\overline{G})j(\overline{G}) = 1$ we obtain that $j(\overline{G}) \in \overline{R}^*$. So $\overline{j(G)} \in \overline{R}^*$. But this implies that $j(G) \in R[x]^*$, which completes the proof.   □

Example 5.2.9 The following statements are equivalent:

(i) $f(x,\zeta)$ has the MZ-property with respect to $ker\ \mathscr{E}$.
(ii) $F = x - H$ is invertible over R.
(iii) For every $g \in R[x]$ with $deg\ g \leq 1$ we have that $gf(x,\zeta)^m \in ker\ \mathscr{E}$ for all large m.

Proof From the Abhyankar–Gurjar formula, taking $u(x) = j(F)^{-1}x_i$, we get

$$\sum_{m\geq 0}\sum_{|a|=m} \frac{1}{a!}\partial^a(H^a(x)x_i) = j(G)G_i.$$

The multinomial formula gives

$$\mathscr{E}(f(x,\zeta)^m x_i) = m! \sum_{|a|=m} \frac{1}{a!} \partial^a (H^a(x)x_i).$$

Suppose that (iii) holds. Then it follows from the two formulas above that $j(G)G_i \in R[x]$ for all i. Since $j(G) \in R[x]^*$, as shown in the proof of Example 5.2.8, we deduce that $G_i \in R[x]$ for every i, i.e., F is invertible over R. So (iii) implies (ii).

Now assume that F is invertible over R. Then $G(x) \in R[x]^n$ and $j(G) \in R[x]^*$. Let $g(x) = x^b \in R[x]$ and take $u(x) = j(F)^{-1}g(x)$ in the Abhyankar–Gurjar formula. Then

$$\sum_{m \geq 0} \sum_{|a|=m} \frac{1}{a!} \partial^a (H^a(x)g(x)) = j(G)g(G).$$

So the left-hand side is a polynomial. It follows that its homogeneous components of large degree are zero. Together with the formula

$$\mathscr{E}(f(x,\zeta)^m g(x)) = m! \sum_{|a|=m} \frac{1}{a!} \partial^a (H^a(x)g(x))$$

this shows that $\mathscr{E}(f(x,\zeta)^m g(x)) = 0$ for large m. In particular for all $b \in \mathbb{N}^n$ we get $\mathscr{E}(f(x,\zeta)^m x^b) = 0$ for all large m, which obviously implies that also $\mathscr{E}(f(x,\zeta)^m x^b \zeta^c) = 0$ for all $b, c \in \mathbb{N}^n$. From this we readily get (i). This shows that (ii) implies (i). Since obviously (i) implies (iii) we are done.   □

Remark 5.2.10 From the two examples above one immediately gets another proof of the fact that the Special Image Conjecture implies the Jacobian Conjecture. But more importantly, it shows us that in the general context of MZ-spaces, the condition that an element a belongs to the radical of some space M can be interpreted as a kind of "generalized Jacobian condition" and similarly that an element a belongs to the strong radical of M can be seen as a "generalized invertibility condition."

The examples above show that it is not at all easy to see if a given subspace of a ring is an MZ-space or not. However, there is an easy property that MZ-spaces share with ideals. This property can often be used to show that a certain space is *not* an MZ-space.

Example 5.2.11 Let M be a left MZ-space of A and $e \in M$ an idempotent, i.e., $e^2 = e$. Then $Ae \subseteq M$. In particular, if $1 \in M$, then $M = A$.

Proof Since $e^m = e \in M$ for all m, it follows that $e \in r(M)$. Since M is a left MZ-space it follows that for every $b \in A$ also $be^m \in M$ for large m, i.e., $be \in M$ for all $b \in A$. So $Ae \subseteq M$. $\qquad\square$

Remark 5.2.12 In Theorem 5.3.1 we show that when A is an algebra over a field k, such that all elements of $r(M)$ are algebraic over k, the converse of Example 5.2.11 holds.

In the next two examples k denotes a field of characteristic zero, although in the first example k can be any commutative ring. Furthermore we let $A := k[x, y]$.

Example 5.2.13 Let $D = \partial_x + y^2 \partial_y$ be a k-derivation on A. Then $Im\ D$ is not an MZ-space of A.

Proof Clearly, $1 = D(x) \in Im\ D$. However $y \notin Im\ D$, since for any $g \in A$ the y-degree of Dg cannot be 1. Now use Example 5.2.11. $\qquad\square$

So the condition $1 \in Im\ D$ is not sufficient to guarantee that $Im\ D$ is an MZ-space of A. However, if we add one condition, namely that the divergence of D equals zero, we obtain a remarkable result:

Example 5.2.14 The following conditions are equivalent:

(i) For every k-derivation D on A such that $1 \in Im\ D$ and div $D = 0$, $Im\ D$ is an MZ-space of A.
(ii) The two-dimensional Jacobian Conjecture is true.

Proof $(i) \Rightarrow (ii)$ Let $F = (f, g) \in k[x, y]^2$ with $det\ JF = 1$ and put $D := g_y \partial_x - g_x \partial_y$. Then $div\ D = 0$ and $1 = det\ JF = Df \in Im\ D$. Since by our hypothesis $Im\ D$ is an MZ-space of A, it follows from Example 5.2.11 that $Im\ D = A$, i.e., D is surjective. Then a result of Stein in [111] (see also [10] or [48]) implies that D is locally nilpotent. Since $D = \frac{\partial}{\partial f}$ it follows from Proposition 2.2.15 in [117] that $ker\ D = ker\frac{\partial}{\partial f} = k[g]$. Since f is a slice of D, we deduce that $k[x, y] = k[g][f]$. So F is invertible over k, i.e., the two-dimensional Jacobian Conjecture is true.

$(ii) \Rightarrow (i)$ Let $d = a\partial_x + b\partial_y$ be a k-derivation on A with $div\ D = 0$ and $1 \in Im\ D$. Since $div\ D = 0$ we get $\partial_x a = \partial_y(-b)$. So by Poincaré's lemma there exists $g \in A$ such that $a = \partial_y g$ and $b = -\partial_x g$. So $D = g_y \partial_x - g_x \partial_y$. Since $1 \in Im\ D$ we get $1 = Df$ for some $f \in A$. Now let $F := (f, g) \in k[x, y]^2$. Then $det\ JF = Df = 1$. Since we assume (ii) it follows that $k[x, y] = k[f, g]$. Hence

$$Im\ D = \frac{\partial}{\partial f}([k[f, g]) = k[f, g] = A$$

In particular $Im\ D$ is an MZ-space of A. $\qquad\square$

Remark 5.2.15 Let $A = k[x, y]$. It is shown in [126, Theorem 3.1] that if $n \leq 2$, then $Im\, D$ is an MZ-space of A for any locally finite k-derivation D on A. The case $n \geq 3$ remains an open problem. On the other hand if A is any commutative k-algebra one easily verifies that if D is a locally nilpotent derivation on A having a slice in A, then $Im\, D = A$ and hence is an MZ-space of A. In fact the same holds for locally finite derivations, see Exercise 6.

To conclude this section we describe some basic properties of the radical, the (left)strong radical, and MZ-spaces. We will use the following notations: $f : A \rightarrow B$ denotes a ring homomorphism sending 1 to 1. Furthermore, we will often denote a subset of A by M and a subset of B by N. The next results are left as exercises to the reader, since they follow straightforward from the definitions.

Proposition 5.2.16

(i) $r(f^{-1}(N)) = f^{-1}(r(N))$.
(ii) $f^{-1}(lsr(N)) \subseteq lsr(f^{-1}(N))$.
(iii) If f is surjective, then $f^{-1}(lsr(N)) = lsr(f^{-1}(N))$.

In the previous proposition we considered the pre-image of a radical. Now we look at images of radicals. As we will see below pre-images are much better behaved than images.

Proposition 5.2.17

(i) $f(r(M)) \subseteq r(f(M))$.
(ii) If f is surjective, then $f(lsr(M)) \subseteq lsr(f(M))$.

The next result describes under what hypothesis on f and M the converses of the inclusions in Proposition 5.2.17 hold:

Proposition 5.2.18 *Let f be surjective and M additive such that $ker(f) \subseteq M$. Then:*

(i) $f^{-1}(r(f(M))) \subseteq r(M)$ and $f^{-1}(lsr(f(M))) \subseteq lsr(M)$.
(ii) $f(r(M)) = r(f(M))$ and $f(lsr(M)) = lsr(f(M))$.

Proof

(i) Let $f(a) \in r(f(M))$. Then there exists N such that $f(a)^m \in f(M)$ for all $m \geq N$. So there exist $\mu_m \in M$ such that $f(a^m) = f(\mu_m)$ for all $m \geq N$. Hence $a^m - \mu_m \in ker(f) \subseteq M$ for all $m \geq N$. Since M is additive it follows that $a^m \in M$ for all $m \geq N$. So $a \in r(M)$. The second part of (i) is proved in a similar way.

(ii) The first equality follows from (i) and Proposition 5.2.17 (i). Finally, to show the second equality in (ii) we only need to show that $lsr(f(M)) \subseteq f(lsr(M))$ (because of Proposition 5.2.17 (ii)). So let $b \in lsr(f(M))$ and write $b = f(a)$ for some $a \in A$. It suffices to show that $a \in lsr(M)$. Therefore choose $c \in A$. We need to show that there exists $N_c \geq 1$ such that $ca^m \in M$ for all $m \geq N_c$. We know that $f(a) \in lsr(f(M))$. So in particular there exists $N_c \geq 1$ such that $f(c)f(a)^m \in f(M)$ for all $m \geq N_c$. But this implies that there exists $\mu_m \in M$ such that $ca^m - \mu_m \in ker(f) \subseteq M$ for all $m \geq N_c$. It follows from the additivity of M that $ca^m \in M$ for all $m \geq N_c$, which completes the proof. $\qquad\square$

Theorem 5.2.19 *Let f be surjective and M additive such that $ker(f) \subseteq M$. Then M is an MZ-space of A if and only if $f(M)$ is an MZ-space of B.*

Proof If M is an MZ-space of A, then $lsr(M) = r(M)$. Hence, applying f and using Proposition 5.2.18 (ii), we get that $f(M)$ is an MZ-space of B. Conversely, assume $f(M)$ is an MZ-space of B. We need to show that $r(M) \subseteq lsr(M)$. So let $a \in r(M)$. Then $f(a) \in f(r(M)) = r(f(M)) \subseteq lsr(f(M))$ (since $f(M)$ is MZ). So $a \in f^{-1}(lsr(f(M))) \subseteq lsr(M)$, by the second part of Proposition 5.2.18 (i). This completes the proof. $\qquad\square$

Corollary 5.2.20 *Let M be an additive subset of A and $I \subseteq A$ a left ideal of A such that $I \subseteq M$. Then M is an MZ-space of A if and only if M/I is an MZ-space of A/I.*

Proof Apply Theorem 5.2.19 to the ring $B := A/I$ and the canonical map $f : A \to A/I$. $\qquad\square$

Proposition 5.2.21 *Let M be an additive subset of A such that $1 \in M$. If $r(M)$ is additive, then $r(M) \subseteq M$. In particular, if $r(M) = A$ then $M = A$.*

Proof If $r(M)$ is not contained in M, there exists $a \in r(M)$ with $a \notin M$. Choose $r \geq 1$ such that $a^r \notin M$ and $a^{r+1}, a^{r+2}, \ldots \in M$. Then in particular $a^r \in r(M)$. Since $1 \in M$, also $1 \in r(M)$. The additivity of $r(M)$ implies that $1 + a^r \in r(M)$. So there exists N such that $(1 + a^r)^N \in M$ and $(1 + a^r)^{N+1} \in M$. So $1 + Na^r \in M$ and $1 + (N + 1)a^r \in M$. Subtracting these elements gives $a^r \in M$, a contradiction. So $r(M) \subseteq M$. Finally, the second statement follows readily from the first. $\qquad\square$

If M is an additive subset of A, then the ideal 0 is contained in M. So we can define I_M as the **largest (left) ideal of A contained in** M. In fact I_M is equal to the sum of all (left) ideals contained in M. The following result, where k is an arbitrary field, will be used in Sect. 5.4.

Proposition 5.2.22 *Let M be a k-linear subspace of $A := k[t]$. Then $sr(M) = r(I_M)$.*

Proof Let $a \in r(I_M)$. Then there exists $N \geq 1$ such that $a^N \in I_M$. Hence for every $b \in A$ and $m \geq N$ we get $ba^m \in I_M \subseteq M$. So $a \in sr(M)$. Conversely, let $a \in sr(M)$. If $a \in k, a \neq 0$, then $1 \in sr(M)$, hence $M = A$. So $I_M = A$ which gives that $a \in r(I_M)$. Now let $a \in A\backslash k$. Then the ring extension $k[a] \subseteq A = k[t]$ is finite, say $a_1, \ldots, a_s$ generate A as a $k[a]$-module. Since $a \in sr(M)$ there exists an N such that $a_i a^m \in M$ for all $1 \leq i \leq s$ and all $m \geq N$. Hence $a_i k[a] a^N \subseteq M$ for all $1 \leq i \leq s$. Since the a_i generate A as a $k[a]$-module it follows that $Aa^N \subseteq M$. So $Aa^N \subseteq I_M$, whence $a \in r(I_M)$.   □

Corollary 5.2.23 *M is an MZ-space of $k[t]$ if and only if $r(M) = r(I_M)$.*

Proof Follows from Proposition 5.2.22 and Definition 5.2.1.   □

Exercises for Sect. 5.2

1. Recall that for a commutative ring R contained in a $\mathbb{Q}$-algebra $\mathfrak{L} : R[x] \to R$ is the R-linear map defined by

$$\mathfrak{L}(x_1^{i_1} \cdots x_n^{i_n}) = i_1! \cdots i_n!$$

   and the Factorial Conjecture, $FC(R, n)$, is the statement: $r(ker\ \mathfrak{L}) \subseteq \mathfrak{n}(R[x])$. Show that $FC(\mathbb{C}, n)$ implies $FC(R, n)$ [Apply the reduction technique used in the proof of Example 5.2.4].

2. Let f be as in (5.2.1). (i) Show that $\mathfrak{L}(f^m) = (m!)^{n+1} u_m$ for all $m \geq 1$.
   (ii) Deduce that $R(k)$ holds for all $1 \leq k \leq n$ if and only if all f of the form (5.2.1) satisfy the strong factorial conjecture.

3. Let $D = \partial_x + y^2 \partial_y - 2yz \partial_z$ on $k[x, y, z]$. Prove that $1 \in Im D$, $div D = 0$ but $Im D$ is not an MZ-space of $k[x, y, z]$. [Show that $y \notin Im D$: if $y = D(\sum_{i=0}^n g_i y^i)$, with $g_i \in k[x, z]$, prove that $z | g_i$ for all $i \geq 1$ and look at the coefficient of y].

4. Let A be a commutative $\mathbb{Q}$-algebra and M an additive subset of A. Let $A[\varepsilon] = A[T]/(T^2)$, where $\varepsilon := \overline{T}$. Prove that the following statements are equivalent:
   (i) M is an MZ-space of A.
   (ii) $r(M) + A\varepsilon \subseteq r(M + \varepsilon M)$.
   (iii) $r(M) + A\varepsilon \subseteq r(A + \varepsilon M)$.

5. (Kumar) Let k be a field and $L : k[t] \to k$ a nonzero k-linear map. Put $a_n = L(t^n)$ and let $a = (a_n)_{n \geq 0}$. We say that a satisfies the **Kumar Condition KC** if for every $d \geq 1$ there exists $N \geq 0$ such that

$$\sum_{i \geq 0} k(a_{N+i}, a_{N+i+1}, \ldots, a_{N+i+d-1}) = k^d.$$

   Show: $sr(ker(L)) = 0$ if and only if a satisfies KC [Use Proposition 5.2.22].

6. Let k be a field of characteristic zero and A a commutative k-algebra. Show that if D is a locally finite k-derivation on A having a slice $s \in A$, then $Im D = A$ [(Nowicki): If

$a \in A$, there exist $n \geq 1$ and $c_i \in k$ such that $D^n a + c_{n-1} D^{n-1} a + \cdots + c_0 a = 0$. Let j be minimal with $c_j \neq 0$. We may assume that $j \geq 1$. Put $b := D^{n-j} a + \cdots + c_j a$ and show that D is locally nilpotent on $R := k[b, Db, \ldots, D^{j-1} b, s]]$.

7. (Zhao) Let A be a commutative ring contained in a $\mathbb{Q}$-algebra and let $\phi : A \to A$ be a ring homomorphism.

   (i) Show that if $\phi^i = \phi^j$ for some $1 \leq i < j$, then $r(Im(1 - \phi)) = r(I)$, where I is the set of $a \in A$ such that $\phi^r(a) = 0$ for some $r \geq 1$.

   (ii) Deduce that $Im(1 - \phi)$ is an MZ-space of A. [For i) put $g(\phi) := \sum_{k=0}^{j-i-1} \phi^k$. Then $\phi^i g(\phi)(1 - \phi) = 0$ implies that $Im(1 - \phi) \subseteq ker\, \phi^i g(\phi)$. So if $a^m \in Im(1 - \phi)$ for all large m deduce that $\phi^i(a), \phi^{i+1}(a), \ldots, \phi^{j-1}(a) \in \mathfrak{n}(A)$ by using Example 5.2.4. Deduce that $a \in r(I)$].

Notes

The result of Example 5.2.5 for the case that R is a domain was first obtained by Zhao and Willems in [140]. For the case $R = \mathbb{C}$ the Examples 5.2.8 and 5.2.9 can essentially be found in the proof of Theorem 3.7 of [137]. Example 5.2.14 was obtained in [126] where it was also shown that the image of every locally finite k-derivation on the polynomial ring in two variables over k is an MZ-space. Proposition 5.2.22 and its corollary can be found as Theorem 2.3 in [125]. The Kumar Condition was communicated by Mohan Kumar to the author during his stay at Washington University in St Louis in 2009.

5.3   Zhao's Idempotency Theorem

In this section k denotes an arbitrary field and A is an associative k-algebra. By M we denote a k-linear subspace of A. As usual, an element $a \in A$ is called **algebraic over k** if there exists a nonzero univariate polynomial $p(t) \in k[t]$ such that $p(a) = 0$. The main result of this section is:

Theorem 5.3.1 (Idempotency Theorem) *Let M be such that all elements of $r(M)$ are algebraic over k. Then M is an MZ-space of A if and only if $Ae \subseteq M$ for all idempotents e which belong to M.*

The proof is based on the following lemma:

Lemma 5.3.2 *If $a \in r(M)$ is algebraic over k there exist $n \geq 1$ and $e \in M$ with $e^2 = e$ such that $a^m \in M$ for all $m \geq n$ and $a^n e = a^n$. If a is not nilpotent, then $e \neq 0$.*

Proof Let $a \in r(M)$. Then for some N $a^m \in M$ for all $m \geq N$. Let $0 \neq p(t) \in k[t]$ with $p(a) = 0$. Then $q(t) := t^N p(t) = t^n h(t)$ with $n \geq N$ and $h(0) \neq 0$. Also $q(a) = 0$. Let $u(t), v(t) \in k[t]$ with

$$u(t)t^n + v(t)h(t) = 1. \tag{5.3.1}$$

Put $e(t) = t^n u(t)$ and $e = e(a)$. Multiply (5.3.1) by t^n and substitute $t = a$. Using $q(a) = 0$ we get $a^n e = a^n$. Multiply (5.3.1) by $e(t)$ and substitute $t = a$. We get $e^2 = e$. Also $e = e(a) \in M$, since $n \geq N$. The last claim follows from $a^n e = a^n$. $\square$

Proof of Theorem 5.3.1 ($\Rightarrow$) is already proved in Example 5.2.11. Conversely, let $a \in r(M)$. Choose e and n as in Lemma 5.3.2. If $b \in A$ and $m \geq n$, then $ba^m = ba^{m-n} \cdot a^n = ba^{m-n}(a^n e) \in Ae \subseteq M$, so M is an MZ-space of A. $\square$

The last line of the proof given above shows that for an MZ-space of A such that all elements of $r(M)$ are algebraic over k the following property holds: if $a \in r(M)$, then there exists $n \geq 1$ such that $Aa^n \subseteq M$. So we get:

Corollary 5.3.3 *If all elements of $r(M)$ are algebraic over k, then M is an MZ-space of A if and only if for every $a \in r(M)$ there exists $n \geq 1$ such that $Aa^n \subseteq M$.*

Proposition 5.3.4 *Let A be commutative and assume that all elements of $r(M)$ are algebraic over k. Then M is an MZ-space of A if and only if $r(M)$ is an ideal in A.*

Proof ($\Rightarrow$) Let $a, b \in r(M)$. By Corollary 5.3.3 there exist n respectively m such that $Aa^n \subseteq M$ and $Ab^m \subseteq M$. It follows that $A(a + b)^{n+m} \subseteq M$ (since A is commutative). Also for every $c \in A$ obviously $A(ca)^n \subseteq Aa^n \subseteq M$. Summarizing: $r(M)$ is an ideal in A.

($\Leftarrow$) Let $e \in M$ be an idempotent. According to Theorem 5.3.1 we must show $Ae \subseteq M$. So if $M_e := \{a \in A \mid ae \in M\}$ and we can show that $M_e = A$ we are done. According to Proposition 5.2.21 it suffices to show that $r(M_e) = A$. So let $b \in A$. Then since $e \in r(M)$ and this is an ideal, also $be \in r(M)$. Hence for all large m we have $b^m e = (be)^m \in M$. But $b^m e \in M$ for large m just means that $b \in r(M_e)$. Since this holds for all $b \in A$, we get $A \subseteq r(M_e)$, which implies the desired equality. $\square$

Remark 5.3.5 If A is not commutative the radical of an MZ-space need not be an ideal, as can be seen from Example 5.2.4. Also the condition that all elements of $r(M)$ are algebraic over k cannot be dropped, even in the commutative case, as can be seen from Exercise 1 in Sect. 5.5.

Example 5.2.4 implies that the set of trace zero matrices over a field k of characteristic zero is a left MZ-space, since its radical is contained in the set of nilpotent matrices. In fact it follows from the next result that any proper *two-sided* MZ-space of $M_n(k)$ has this property, since $M_n(k)$ is a simple ring and obviously all its elements are algebraic over k.

Proposition 5.3.6 *Let A be a simple ring and M a proper subspace of A such that all elements of $r(M)$ are algebraic over k. Then M is a two-sided MZ-space of A if and only if $r(M) \subseteq \mathfrak{n}(A)$.*

Proof By Example 5.2.3 it remains to prove ($\Rightarrow$). Suppose that $r(M)$ is not contained in $\mathfrak{n}(A)$. Then there exists $a \in r(M)$ which is not nilpotent. Hence it follows from the second part of Lemma 5.3.2 that M contains a nonzero idempotent e. So $e \in r(M)$. Consequently, for every $b, c \in A$ we have $be^m c \in M$ for all large m. So $bec \in M$ for all $b, c \in A$. This means that M contains a nonzero two-sided ideal. Since A is simple it follows that $A \subseteq M$, i.e., $M = A$, contradicting that M is a proper subspace of A. So $r(M) \subseteq \mathfrak{n}(A)$. $\square$

In Sect. 5.8 we will give a complete description of all codimension one (left) MZ-spaces of rings which are finite products of matrix rings over k.

Exercises for Sect. 5.3

1. Let k be a field of characteristic zero and $\Lambda \subseteq k^*$ a finite set. For each $\lambda \in \Lambda$ let $S_\lambda : k[t] \to k$ be the substitution homomorphism sending t to λ. Put $D := t\partial_t$. Show that
   (i) $(S_\lambda \circ D^i)((t - \lambda)^j k[t]) = 0$, if $j > i$.
   (ii) $(S_\lambda \circ D^i)((t - \lambda)^i u) \neq 0$, if $u(\lambda) \neq 0$.
2. Notations as in 1. For each $\lambda \in \Lambda$ let d_λ be a non-negative integer and $P_\lambda(T) \in k[T]$ a polynomial of degree d_λ. Let $L : k[t] \to k$ be the k-linear map defined by

$$L = \sum_{\lambda \in \Lambda} S_\lambda \circ P_\lambda(D).$$

   Put $V = ker\, L$. Show that $I_V = k[t] \prod_{\lambda \in \Lambda} (t - \lambda)^{d_\lambda + 1}$ [Use Exercise 1].
3. Notations as in 2. By the Chinese Remainder Theorem we have an isomorphism

$$\phi : k[t]/I_V \to \prod_{\lambda \in \Lambda} k[t]/k[t](t - \lambda)^{d_\lambda + 1} =: B$$

   given by $\phi(g + I_V) = \left(g + k[t](t - \lambda)^{d_\lambda + 1}\right)_{\lambda \in \Lambda}$.
   (i) Show that the elements $e_\lambda = (0, \ldots, 0, 1, 0, \ldots, 0)$, with 1 at the component with index λ, form an **orthogonal basis of idempotents of B**, i.e., each e_λ is an idempotent of B, $e_\lambda e_\mu = 0$ if $\lambda \neq \mu$ and each nonzero idempotent of B is of the form $\sum_{\lambda \in \Lambda'} e_\lambda$ for some non-empty subset Λ' of Λ.
   (ii) Let $\phi(g_\lambda + I_V) = e_\lambda$. Deduce that the $g_\lambda + I_V$ form an orthogonal basis of idempotents of $k[t]/I_V$.

4. Let R be a commutative ring having an orthogonal basis E of idempotents. Show that if M is an MZ-space of R which does not contain any element of E, then 0 is the only idempotent of R contained in M.

5. Notations as above. Show that $V = \ker L$ is an MZ-space of $k[t]$ if and only if for each non-empty open subset Λ' of Λ we have that $\sum_{\lambda \in \Lambda'} L(g_\lambda) \neq 0$ [Use Zhao's idempotency theorem and Exercises 3 and 4].

6. Notations as above
   (i) Show that $L(g_\lambda) = P_\lambda(0)$ [Use Exercise 1 and the fact that $g_\lambda \equiv 1 \bmod (t-\lambda)^{d_\lambda+1}$ and $g_\lambda \equiv 0 \bmod (t - \mu)^{d_\mu+1}$ if $\lambda \neq \mu$].
   (ii) Deduce from Exercise 5 that V is an MZ-space of $k[t]$ if and only if for every non-empty subset Λ' of Λ we have $\sum_{\lambda \in \Lambda'} P_\lambda(0) \neq 0$.

7. Let k be a field of characteristic zero, A a commutative k-algebra, and $D : A \to A$ a k-derivation. Show that if $a \in A$ is algebraic over k and $e := Da$ is an idempotent, then $e = 0$ [Show that $De = 0$].

8. Let k be a field of characteristic zero and A a commutative k-algebra such that all its elements are algebraic over k. Let $D : A \to A$ be a k-derivation and M a k-linear subspace of A. Show that $D(M)$ is an MZ-space of A.

9. Let A be a commutative ring, S a multiplicatively closed subset of A, and B an A-algebra. An A-submodule M of B is called S-**saturated** if for all $s \in S$ and all $b \in B$ the relation $sb \in M$ implies that $b \in M$.
   Let M be an S-saturated A-submodule of B.
   (i) Show that $S^{-1}r(M) = r(S^{-1}M)$ and $S^{-1}sr(M) = sr(S^{-1}M)$, where for a subset V of B, $S^{-1}V$ is the set of all elements v/s, with $v \in V$ and $s \in S$.
   (ii) Show that M is an MZ-space of B if and only if $S^{-1}M$ is an MZ-space of $S^{-1}B$.

10. (A generalization of Theorem 5.3.1) Let A be a domain, $S = A \setminus \{0\}$, and B an A-algebra. Let M be an S-saturated A-submodule of B such that all elements of $r(M)$ are algebraic over A. Show that the following two statements are equivalent:
    (i) M is an MZ-space of B.
    (ii) For every S-idempotent $e \in B$ with $e \in M$ we have that for every $b \in B$ there exists an $s \in S$ such that $sbe \in M$ ($e \in B$ is called an S-**idempotent** if there exists an $s \in S$ such that $e^2 = se$) [Use Theorem 5.3.1].

Notes

The Idempotency theorem is Theorem 4.2 of [139]. The proof given here is a somewhat simplified version of Zhao's original proof. The first six exercises above are taken from the paper [121], where a complete description is given of all Mathieu–Zhao spaces of $k[t]$ having a nonzero strong radical. This result was first obtained in the Master's thesis [95] of Simeon Nieman. The proof of the main result given there is much more involved than the one given in [121]. In 2015 the results of [121] were extended to the n-variable case in the Master's thesis [130] of Loes van Hove. A much simplified proof of her main results can be found in [122]: in that paper we describe all Mathieu–Zhao spaces of $k[x_1, \ldots, x_n]$ which contains an ideal of finite codimension, where k is an algebraically closed field of characteristic zero. Furthermore we give an algorithm to decide if a subspace of the form $I + kv_1 + \cdots + kv_r$ is a Mathieu–Zhao space, in case the ideal I has finite codimension.

5.4   Orthogonal Polynomials and MZ-Spaces

An important class of MZ-spaces is formed by several subspaces of polynomial rings over the complex numbers. More precisely, the main conjecture which relates orthogonal polynomials and MZ-spaces asserts the following: if $\{u_a\}_{a \in \mathbb{N}^n}$ is a set of **classical orthogonal polynomials** of $\mathbb{C}[x] := \mathbb{C}[x_1, \ldots, x_n]$, then the codimension one subspace of $\mathbb{C}[x]$ generated by the $\{u_a\}$ with $|a| > 0$ forms an MZ-space of $\mathbb{C}[x]$. We will show that if this conjecture holds for the set of **Hermite polynomials** (in several variables) the Special Image Conjecture holds as well. This in turn implies the Generalized Vanishing Conjecture and hence the Jacobian Conjecture.

We prove the above conjecture for various classes of classical orthogonal polynomials in one variable. In particular for the Laguerre polynomials. From this fact we deduce in Sect. 5.7 that the Special Image Conjecture in one variable is true.

For convenience of the reader we first recall some basic facts concerning orthogonal polynomials (for more information the reader is referred to [41] and [114]).

Orthogonal Polynomials

Let B be a non-empty open subset of $\mathbb{R}^n$ and w a so-called **weight function** on B, i.e., it is strictly positive on B and its integral over this set is finite and positive. To such a function one can associate a **Hermitian inner product** on the n-dimensional polynomial ring $\mathbb{C}[x]$ by defining

$$\langle f, g \rangle = \int_B f(x)\overline{g(x)}w(x)dx.$$

A set of polynomials u_a, where $a = (a_1, \ldots, a_n)$ runs through $\mathbb{N}^n$, is called **orthogonal over B** with respect to the weight function w, if they form an orthogonal basis of $\mathbb{C}[x]$ with respect to the associated inner product described above and satisfy the additional condition that the degree of each polynomial u_a is equal to $|a|$, the sum of all a_i.

A standard way to construct orthogonal polynomials in one variable is to apply the Gram-Schmidt process to the basis $\{1, x, x^2, \ldots\}$. Making special choices for B and w gives the following so-called **classical orthogonal polynomials**.

1. **Hermite polynomials**: $B = \mathbb{R}$, $w(x) = e^{-x^2}$.
2. **Laguerre polynomials**: $B = (0, \infty)$, $w(x) = x^\alpha e^{-x}$, with $\alpha > -1$.

3. **Jacobi polynomials**: $B = (-1, 1)$, $w(x) = (1 - x)^\alpha (1 + x)^\beta$, with $\alpha, \beta > -1$. If both parameters are zero, i.e., $w = 1$, the polynomials are called **Legendre polynomials**.

From univariate orthogonal polynomials one can construct orthogonal polynomials in dimension n as follows: for each $1 \le i \le n$ choose an open subset B_i of $\mathbb{R}$ and a weight function w_i on B_i. Let $u_{i,m}$ with $m \ge 0$ be an orthogonal set of univariate polynomials with respect to B_i and w_i.

Then $B = B_1 \times \cdots \times B_n$ is an open subset of $\mathbb{R}^n$ and w defined by $w(x) = w_1(x_1) \cdots w_n(x_n)$ is a weight function on B, where $x = (x_1, \ldots, x_n)$. Furthermore one easily verifies that the polynomials

$$u_a(x) = u_{1,a_1}(x_1) \cdots u_{n,a_n}(x_n),$$

where $a = (a_1, \ldots, a_n)$ runs through $\mathbb{N}^n$, form an orthogonal set of polynomials over B with respect to the weight function w.

The multivariate orthogonal polynomials obtained from Hermite polynomials will again be called Hermite polynomials. Similarly we get multivariate Laguerre and Jacobi polynomials. These polynomials we call the **classical (multivariate) orthogonal polynomials**.

In spite of the fact that hundreds of papers are concerned with orthogonal polynomials, the following conjecture, due to Zhao, seems to be almost completely open:

Classical Orthogonal Polynomial Conjecture COPC(n) *Let $\{u_a\}_{a \in \mathbb{N}^n}$ be a set of classical orthogonal polynomials in $\mathbb{C}[x]$. Then $M := \sum_{|a| > 0} \mathbb{C}u_a$ is an MZ-space of $\mathbb{C}[x]$.*

Let M and its orthogonal polynomials be defined by B, w, and $\langle, \rangle$. Then

Proposition 5.4.1 $M = \{f \in \mathbb{C}[x] \mid \int_B f(x)w(x)\, dx = 0\}$.

Proof Let $f \in \mathbb{C}[x]$. Write f on the basis $\{u_a\}$, say $f = \sum_a f_a u_a$, with f_a in $\mathbb{C}$. It follows that $f \in M$ if and only if $f_0 = 0$. Since the $\{u_a\}$ form an orthogonal basis of $\mathbb{C}[x]$, we get

$$f_0 \langle u_0, u_0 \rangle = \langle f, u_0 \rangle = \int_B f\, \overline{u_0}\, w\, dx = \overline{u_0} \int_B f w\, dx.$$

So $f \in M$ if and only if $\int_B f w\, dx = 0$, since $u_0 \in \mathbb{C}^*$. $\square$

Corollary 5.4.2 *With the notations as above: COPC(n) for the set of orthogonal polynomials defined by B, w, and $\langle , \rangle$ is equivalent to*

$$\{f \in \mathbb{C}[x] \mid \int_B f(x)w(x)\,dx = 0\}$$

is an MZ-space of $\mathbb{C}[x]$.

Consequently, by restricting to the **Hermite polynomials**, i.e., the set of orthogonal polynomials where $B = \mathbb{R}^n$ and the weight function is given by $w(x_1, \ldots, x_n) = e^{-(x_1^2 + \cdots + x_n^2)}$, we obtain the following special case of COPC(n), which we call the **Gaussian Moments Conjecture:**

Gaussian Moments Conjecture (GMC(n))
$\{f \in \mathbb{C}[x] \mid \int_{\mathbb{R}^n} f(x)e^{-(x_1^2 + \cdots + x_n^2)}\,dx = 0\}$ is an MZ-space of $\mathbb{C}[x]$.

The importance of this conjecture comes from the following theorem.

Theorem 5.4.3 (Derksen, van den Essen, Zhao) *If GMC(2n) holds, then SIC(n) holds.*

To prove this result we define the $\mathbb{C}$-linear map

$$\mathscr{F}_n : \mathbb{C}[x, \zeta] \to \mathbb{C} \text{ by } \mathscr{F}_n(f) = \mathscr{E}_n(f) \mid_{x=0},$$

where $\mathscr{E}_n$ is as defined in Sect. 5.2. Theorem 5.4.3 immediately follows from the next two results:

Proposition 5.4.4 *If ker $\mathscr{F}_n$ is an MZ-space of $\mathbb{C}[x, \zeta]$, then ker $\mathscr{E}_n$ is an MZ-space of $\mathbb{C}[x, \zeta]$, i.e., SIC(n) holds.*

Proposition 5.4.5 *GMC(2n) is true if and only if ker $\mathscr{F}_n$ is an MZ-space of $\mathbb{C}[x, \zeta]$.*

Proof of Proposition 5.4.4 Observe that $(\partial^a x^b)_{|x=\alpha} = \partial^a (x + \alpha)^b_{|x=0}$ for all $\alpha \in \mathbb{C}^n$ and all $a, b \in \mathbb{N}^n$. It follows that for all $f(x, \zeta) \in \mathbb{C}[x, \zeta]$

$$\mathscr{E}_n(f(x, \zeta))_{|x=\alpha} = \mathscr{E}_n(f(x + \alpha, \zeta))_{|x=0} = \mathscr{F}_n(f(x + \alpha, \zeta)). \qquad (5.4.1)$$

Now suppose that ker$\mathscr{F}_n$ is an MZ-space and assume that $P^m \in$ ker $\mathscr{E}_n$ for all $m \geq 1$. Then for each $\alpha \in \mathbb{C}^n$ we have

$$0 = \mathscr{E}_n(P^m(x, \zeta))_{|x=\alpha} = \mathscr{E}_n(P^m(x + \alpha, \zeta))_{|x=0} = \mathscr{F}_n(P^m(x + \alpha, \zeta)).$$

So $P^m(x + \alpha, \zeta) \in ker \, \mathscr{F}_n$ for all $m \geq 1$. Now let $Q \in \mathbb{C}[x, \zeta]$. Since $ker \mathscr{F}_n$ is an MZ-space it follows that there exists some N such that for all $m \geq N$ $\mathscr{F}_n(Q(x + \alpha, \zeta)P^m(x + \alpha, \zeta)) = 0$. Hence by (5.4.1)

$$\mathscr{E}_n(Q(x, \zeta)P^m(x, \zeta))\big|_{x=\alpha} = 0 \text{ for all } m \geq N. \tag{5.4.2}$$

Let $V_N \subseteq \mathbb{C}^n$ be the zero set of the polynomials $\mathscr{E}_n(Q(x, \zeta)P^m(x, \zeta))$ with $m \geq N$. Since for each $\alpha \in \mathbb{C}^n$ there exists an N such that (5.4.2) holds, it follows that $\bigcup_{N=1}^\infty V_N = \mathbb{C}^n$. Hence $V_N = \mathbb{C}^n$ for some N, since $\mathbb{C}^n$ cannot be a countable union of proper closed subsets. It follows that we have $\mathscr{E}_n(Q(x, \zeta)P^m(x, \zeta)) = 0$ for all $m \geq N$. $\qquad\square$

In order to prove Proposition 5.4.5 we first make some observations. The first one is that $\mathscr{F}_n(\zeta^\alpha x^\beta) = 0$, if $\alpha \neq \beta$ and is equal to $\alpha!$, if $\alpha = \beta$. Furthermore we define the $\mathbb{C}$-linear map L from the polynomial ring $\mathbb{C}[x, y] := \mathbb{C}[x_1, \ldots, x_n, y_1, \ldots, y_n]$ to $\mathbb{C}$ by

$$L(f(x, y)) = \int_{\mathbb{R}^{2n}} e^{-(\sum x_j^2 + \sum y_j^2)} f(x, y) \, dx dy.$$

Lemma 5.4.6 $L(f(x + iy, x - iy)) = \pi^n \mathscr{F}_n(f(x, \zeta))$ for all $f(x, \zeta) \in \mathbb{C}[x, \zeta]$.

Proof Since both L and $\mathscr{F}_n$ are $\mathbb{C}$-linear, it suffices to prove this equality for the case $f(x, \zeta) = \zeta^\alpha x^\beta$. We have

$$L((x - iy)^\alpha (x + iy)^\beta) = \int_\mathbb{R} \cdots \int_\mathbb{R} e^{-\sum_j (x_j^2 + y_j^2)} (x - iy)^\alpha (x + iy)^\beta \, dx dy$$

$$= \prod_{j=1}^n \Big[\int_\mathbb{R} \int_\mathbb{R} e^{-(x_j^2 + y_j^2)} (x_j - iy_j)^{\alpha_j} (x + iy)^{\beta_j} \, dx_j dy_j \Big].$$

Introducing polar coordinates, $x_j = r_j cos(t_j)$, $y_j = r_j sin(t_j)$, gives

$$= \prod_{j=1}^n \Big[\int_{r_j=0}^\infty \int_{t_j=0}^{2\pi} e^{-r_j^2} r_j^{\alpha_j + \beta_j} e^{i(\beta_j - \alpha_j)t_j} r_j \, dt_j dr_j \Big].$$

If one of the factors in this product equals zero, the whole product equals zero. This is the case if there exists some j such that $\alpha_j \neq \beta_j$. This implies that $L((x - iy)^\alpha (x + iy)^\beta) = 0$ if $\alpha \neq \beta$. Since, as observed above, also $\mathscr{F}_n(\zeta^\alpha x^\beta) = 0$ if $\alpha = \beta$, this shows that in this case the statement of the lemma is correct for $f(x, \zeta) = \zeta^\alpha x^\beta$. Furthermore, if $\beta = \alpha$ we get

$$L((x - iy)^\alpha (x + iy)^\alpha) = \prod_{j=1}^n \Big[2\pi \int_{r_j=0}^\infty e^{-r_j^2} r_j^{2\alpha_j} r_j \, dr_j \Big].$$

Making the coordinate change $r^2 = s$, we get that for all $a \in \mathbb{N}$

$$\int_{r=0}^{\infty} e^{-r^2} r^{2a} r\, dr = \frac{1}{2} \int_{s=0}^{\infty} e^{-s} s^a\, ds = \frac{1}{2} a!.$$

So

$$L((x - iy)^{\alpha} (x + iy)^{\alpha}) = (2\pi)^n (\frac{1}{2})^n \alpha! = \pi^n \alpha!.$$

Since as observed above $\mathscr{F}_n(\zeta^{\alpha} x^{\alpha}) = \alpha!$, this implies the lemma.   □

Proof of Proposition 5.4.5 Let $\phi : \mathbb{C}[x, \zeta] \to \mathbb{C}[x, y]$ be given by $\phi(f(x, \zeta)) = f(x + iy, x - iy)$. Then ϕ is an isomorphism. From the lemma it follows that $\phi(\ker \mathscr{F}_n) = \ker L$. So $\ker L$ is an MZ-space of $\mathbb{C}[x, y]$ if and only if $\ker \mathscr{F}_n$ is an MZ-space of $\mathbb{C}[x, \zeta]$.   □

Before we restrict our attention to the one-dimensional case let us make the following remark: it was conjectured in [138] that Corollary 5.4.2 also holds for *arbitrary* open $B \subseteq \mathbb{R}^n$ and weight functions w on B. However, the following example shows that this conjecture is false:

Example 5.4.7 Let H be the compact half-disk consisting of the $z \in \mathbb{C}$ such that $z \leq 1$ and $Re\, z \geq 0$, $f(x, y) = (x + iy)^2$ and $g(x, y) = x + iy$. Then $\int_B f(x, y)^m\, dxdy = 0$ for all $m \geq 1$, but $\int_B g(x, y) f(x, y)^m\, dxdy \neq 0$ for all $m \geq 1$.

Proof $\int_B f(x, y)^m\, dxdy = \int_{r=0}^1 \int_{t=0}^{\pi} r^{2m} e^{2mit} r\, dt = 0$ for all $m \geq 1$ and $\int_B g(x, y) f(x, y)^m\, dxdy = \int_{r=0}^1 \int_{t=0}^{\pi} r^{2m+1} e^{(2m+1)it} r\, dt \neq 0$ if $m \geq 1$.   □

However, in dimension one the following conjecture is still open:

Integral Conjecture *Let $B \subset \mathbb{R}$ be an open subset and σ a positive measure such that $\int_B g(t) d\sigma$ is finite for each $g \in \mathbb{C}[t]$. Set*

$$M_B(\sigma) := \left\{ f \in \mathbb{C}[t] \,\middle|\, \int_B f\, d\sigma = 0 \right\}.$$

Then $M_B(\sigma)$ is an MZ-space of $\mathbb{C}[t]$.

Before we investigate this conjecture we first prove a very special case of it, which even holds in the n-dimensional case. More precisely:

Proposition 5.4.8 *Let $B \subseteq \mathbb{R}^n$ be an open subset and $S := \{b_1, \ldots, b_d\}$ a finite subset of B. If σ is an atomic measure supported at S (i.e., $\sigma(b_i) > 0$ for all i and for every* measurable set $U \subseteq B$ we have $\sigma(U) = \sum_{u \in S \cap U} \sigma(u))$, then $M_B(\sigma)$ is an MZ-space of $\mathbb{C}[x]$.

Proof Let $f \in \mathbb{C}[x]$. Then $f \in M_B(\sigma)$ if and only if

$$\sum_i f(b_i)\sigma(b_i) = 0. \tag{5.4.3}$$

So, if $f \in \mathbb{C}[x]$ satisfies $f^m \in M_B(\sigma)$ for all $m \geq 1$, then

$$\sum_i f^m(b_i)\sigma(b_i) = 0 \text{ for all } m \geq 1. \tag{5.4.4}$$

If $f(b_i) = 0$ for all i, then for any $g \in \mathbb{C}[x]$ and all $m \geq 1$ also gf^m satisfies equation (5.4.3). So $gf^m \in M_B(\sigma)$. Therefore assume that not all $f(b_i)$ are zero and let $\{c_1, \ldots, c_s\}$ be the set of all *distinct* nonzero values which f attains on S. Furthermore, for any $1 \leq j \leq s$ let $S_j := S \cap f^{-1}(c_j)$. Then Eq. (5.4.4) becomes

$$0 = \sum_i f^m(b_i)\sigma(b_i) = \sum_{j=1}^s c_j^m \sum_{b \in S_j} \sigma(b).$$

Since these equalities hold for all $m \geq 1$ it follows, using a Vandermonde matrix, that for each j the expression $\sum_{b \in S_j} \sigma(b)$ is equal to 0, which contradicts the fact that $\sigma(b_i) > 0$ for each i.   □

Strong Integral Conjecture *With the same notations as in the Integral Conjecture assume that σ is not an atomic measure supported at finitely many points. Then $r(M_B(\sigma)) = \{0\}$.*

In other words, this conjecture claims that when the measure σ is not an atomic measure supported at finitely many points, the only polynomial f satisfying $\int_B f^m d\sigma = 0$ for all $m \geq 1$ is the zero polynomial.

Theorem 5.4.9 *The Strong Integral Conjecture and the Integral Conjecture are equivalent.*

Proof Assume that the Strong Integral Conjecture holds. Note first that if σ is an atomic measure supported at finitely many points, then by Proposition 5.4.8 the Integral Conjecture holds. When σ is not an atomic measure supported at finitely many points the hypothesis implies that $r(M_B(\sigma)) = \{0\}$. Then it follows from Example 5.2.3 that $M_B(\sigma)$ is an MZ-space of $\mathbb{C}[t]$. So the Integral Conjecture holds. Conversely, assume that the Integral Conjecture holds and put $M = M_B(\sigma)$. We claim that the largest ideal I_M of $\mathbb{C}[t]$ contained in M is equal to $\{0\}$. It then follows from Proposition 5.2.22 that

$sr(M) = r(0) = \{0\}$ ($\mathbb{C}[t]$ has no nonzero nilpotent elements). Since by hypothesis M is an MZ-space of $\mathbb{C}[t]$ we get $r(M) = sr(M) = \{0\}$, as desired. To prove the claim let us assume the contrary and let $0 \neq f \in I_M$. Then $g := \bar{f} f \in I_M \subseteq M$ (where $\bar{f}$ denotes the complex conjugate of f), whence $g \in M$. Then, by the definition of M, the integral of g over B is equal to zero. On the other hand, since g is continuous and positive over B (except at the finitely many zeros of f in B) and σ is not an atomic measure supported at finitely many points, the integral of g over B is positive, which is a contradiction.   □

To conclude this section we describe a technique, due to Mitya Boyarchenko (unpublished), which allows us to prove that certain subspaces of polynomial rings are MZ-spaces (see also [46]). His method is based on two ingredients. The first one, which is explained in Proposition 5.4.10 below, makes clever use of the Nullstellensatz. The second one uses p-adic absolute values on number fields. We briefly recall some properties of these absolute values below.

Proposition 5.4.10 *Let $L : \mathbb{C}[x] \to \mathbb{C}$ be a $\mathbb{C}$-linear map such that $L(x^a) \in \overline{\mathbb{Q}}$ for all $a \in \mathbb{N}^n$. If there exist $0 \neq f \in \mathbb{C}[x]$ and $N \geq 1$ such that $L(f^m) = 0$ for all $m \geq N$, then there exists $0 \neq F \in \overline{\mathbb{Q}}[x]$ such that $L(F^m) = 0$ for all $m \geq N$.*

Proof   Let $d := \deg f$. Since $f \neq 0$, there exists $a_0 \in \mathbb{N}^n$ with $|a_0| = d$ such that the coefficient of x^{a_0} in f is nonzero. We may assume that this coefficient equals 1. So we can write

$$f = x^{a_0} + \sum_{|a| \leq d, a \neq a_0} c_a x^a.$$

Let $\mathbb{Z}[C]$ be the polynomial ring over $\mathbb{Z}$ in the variables C_a with $|a| \leq d$ and $a \neq a_0$. For each $m \geq N$ we have

$$f^m = \sum_b g_b^{(m)}(c) x^b, \tag{5.4.5}$$

where $c = \{c_a\}_{|a| \leq d, a \neq a_0}$ and $g_b^{(m)}(C) \in \mathbb{Z}[C]$ for all $b \in \mathbb{N}^n$ with $|b| \leq md$. Apply L to (5.4.5). Then it follows from the hypothesis that

$$0 = L(f^m) = \sum_b g_b^{(m)}(c) L(x^b) \text{ for all } m \geq N. \tag{5.4.6}$$

Now define for each $m \geq N$ the polynomial

$$P_m(C) := \sum_b g_b^{(m)}(C) L(x^b).$$

So $P_m(C) \in \overline{\mathbb{Q}}[C]$ for all $m \geq N$ (because $L(x^b) \in \overline{\mathbb{Q}}$).

Since by (5.4.6) $P_m(c) = 0$ for all $m \geq N$ it follows that $1 \notin J$, the ideal in $\overline{\mathbb{Q}}[C]$ generated by the $P_m(C)$ with $m \geq N$. Then the Nullstellensatz implies that there exists $u = \{u_a\}_{|a| \leq d, a \neq a_0} \in \overline{\mathbb{Q}}^M$, where M is the number of monomials x^a, with $|a| \leq d$ and $a \neq a_0$, such that $P_m(u) = 0$ for all $m \geq N$. But this means that $0 \neq F := x^{a_0} + \sum_{|a| \leq d, a \neq a_0} u_a x^a \in \overline{\mathbb{Q}}[x]$ and that $L(F^m) = 0$ for all $m \geq N$, which completes the proof.

$\square$

Before we describe the second ingredient of Boyarchenko's method, we briefly recall some results from number theory.

Let k be a field. An **absolute value** on k is a map $|.| : k \to \mathbb{R}$ satisfying the following properties: $|x| \geq 0$ for all $x \in k$ and $|x| = 0$ if and only if $x = 0$, $|xy| = |x||y|$, and $|x + y| \leq |x| + |y|$ for all $x, y \in k$. If $|x + y| \leq max(|x|, |y|)$ for all $x, y \in k$ the absolute value is called **non-Archimedean**. An example of an absolute value is the trivial one, defined by $|0| = 0$ and $|x| = 1$ for all $x \in k^*$. On $\mathbb{Q}$ one can completely describe all absolute values. First of all there is the usual absolute value defined by $|x| = x$ if $x \geq 0$ and $|x| = -x$ if $x < 0$. Furthermore, for each prime number p we have a non-Archimedean absolute value, denoted $|.|_p$, defined by

$$\left|\frac{a}{b}p^n\right|_p = p^{-n},$$

where a and b are integers which have no factor p. This absolute value is called the p-**adic absolute value** on $\mathbb{Q}$. It is a classical result due to Ostrowski which asserts that every nontrivial absolute value on $\mathbb{Q}$ is **equivalent** to one of the above (two absolute values $|.|_1$ and $|.|_2$ on a field k are called equivalent if there exists $c > 0$ such that $|x|_2 = |x|_1^c$ for all $x \in k$).

Suppose now that $\mathbb{Q} \subseteq K$ is a finite field extension, i.e., K is a **number field**. Let $|.|_p$ be the p-adic absolute value on $\mathbb{Q}$. Then $|.|_p$ can be extended to an absolute value on K, which we again denote by $|.|_p$. There may exist several extensions of $|.|_p$ to K. We choose one such an extension for each p. Then the following result plays a crucial role in the arguments below.

Lemma 5.4.11 *Let $a \in K$. Then there exists $N \geq 1$ such that $|a|_p \leq 1$ for all $p > N$.*

Proof If $a = 0$ we are done. So let $a \neq 0$. Since $\mathbb{Q} \subseteq K$ is finite, a is algebraic over $\mathbb{Q}$. So there exist $q_i \in \mathbb{Q}$ and $d \geq 1$ such that

$$a^d + q_{d-1}a^{d-1} + \cdots + q_1 a + q_0 = 0.$$

Since the q_i have only a finite number of prime factors, there exists $N \geq 1$ such that $|q_i|_p = 1$ for all $q_i \neq 0$ and all $p > N$. So $|q_i|_p \leq 1$ for all i and all $p > N$. For such a p we get

$$|a|_p^d = |a^d|_p = |q_{d-1}a^{d-1} + \cdots + q_1 a + q_0|_p \leq \max_i |q_i a^i|_p \leq \max_{1 \leq i \leq d-1} |a|_p^i.$$

Hence $|a|_p \leq 1$, which completes the proof. $\square$

To illustrate how the results above can be used to show that certain spaces are MZ-spaces, we will prove that the spaces described in Examples 5.2.6 and 5.2.7 are MZ-spaces.

Proposition 5.4.12 *Let* $L : \mathbb{C}[t] \to \mathbb{C}$ *be defined by* $L(f(t)) = \int_0^1 f(t)\,dt$. *Then* $r(\ker L) = \{0\}$.

Proof Suppose $r(\ker L) \neq 0$. Then there exists a monic $f \in \mathbb{C}[t]$ and $N \geq 1$ such that $\int_0^1 f(t)^m\,dt = 0$ for all $m \geq N$. Hence $d := \deg_t f(t) \geq 1$. By Proposition 5.4.10 we may assume that $f \in \overline{\mathbb{Q}}[t]$. Write $f = t^d + \sum_{i=0}^{d-1} c_i t^i$. Then

$$f^m = t^{md} + \sum_{i=0}^{md-1} g_i^{(m)}(c)\, t^i, \tag{5.4.7}$$

where $c = (c_0, \ldots, c_{d-1})$ and $g_i^{(m)}(C_0, \ldots, C_{d-1}) \in \mathbb{Z}[C_0, \ldots, C_{d-1}]$, the polynomial ring in the variables C_j over $\mathbb{Z}$. Applying the linear map L to Eq. (5.4.7) and using that $L(t^i) = \frac{1}{i+1}$ we obtain that

$$0 = L(f^m) = \frac{1}{md+1} + \sum_{i=0}^{md-1} g_i^{(m)}(c)\frac{1}{i+1} \quad \text{for all } m \geq N. \tag{5.4.8}$$

By Dirichlet's prime number theorem there exist infinitely many prime numbers of the form $md+1$. Since all c_j are algebraic over $\mathbb{Q}$, there exists a number field K which contains all the c_j. Now choose $m \geq N$ such that $p = md + 1$ is prime and furthermore large enough to guarantee that $|c_j|_p \leq 1$ for all j (using Lemma 5.4.11). Then $|g_i^{(m)}(c)|_p \leq 1$ for all i. Also observe that for all $0 \leq i \leq md - 1$ we have that $|\frac{1}{i+1}|_p = 1$ (since $1 \leq i + 1 \leq md < p$ and hence p does not divide $i + 1$). Consequently

$$\left| \sum_{i=0}^{md-1} g_i^{(m)}(c)\frac{1}{i+1} \right|_p \leq 1. \tag{5.4.9}$$

Finally look at Eq. (5.4.8) and observe that $|\frac{1}{md+1}|_p = |\frac{1}{p}|_p = p > 1$, which contradicts (5.4.9). $\square$

Proposition 5.4.13 *Let $L : \mathbb{C}[t] \to \mathbb{C}$ be defined by $L(f(t)) = \int_0^\infty f(t)e^{-t}\, dt$. Then $r(\ker L) = \{0\}$.*

Proof Suppose that $r(\ker L) \neq 0$. Then there exist $0 \neq f \in \mathbb{C}[t]$ and $N \geq 1$ such that $\int_0^\infty f(t)^m e^{-t}\, dt = 0$ for all $m \geq N$. By Proposition 5.4.10 we may assume that $f \in \overline{\mathbb{Q}}[t]$. We may also assume that f is of the form

$$f = t^r + c_{r+1} t^{r+1} + \cdots + c_{r+d} t^{r+d}$$

with $d \geq 1$. Then for each $m \geq N$ we can write

$$f^m = t^{rm} + \sum_{i=1}^{md} g_i^{(m)}(c) t^{rm+i}, \tag{5.4.10}$$

where $c = (c_{r+1}, \dots, c_{r+d})$. Then, applying L to (5.4.10) and using that $L(t^i) = \int_0^\infty t^i\, e^{-t}\, dt = i!$, we obtain

$$0 = (rm)! + \sum_{i=1}^{md} g_i^{(m)}(c)(rm+i)! \tag{5.4.11}$$

Again we know that all c_j belong to some number field K. Divide the relation in (5.4.11) by $(rm)!$ and observe that for each $1 \leq i \leq md$ there exists a positive integer n_i such that $\frac{(rm+i)!}{(rm)!} = (rm+1)n_i$. So from Eq. (5.4.11) we deduce

$$-1 = \sum_{i=1}^{md} g_i^{(m)}(c)(rm+1)n_i. \tag{5.4.12}$$

Now again choose m large enough such that $rm + 1 = p$ is prime and $|g_i^{(m)}(c)|_p \leq 1$ for all i. Then it follows that the p-adic absolute value of the right-hand side of (5.4.12) is strictly smaller than 1 (since $|rm + 1|_p = |p|_p = \frac{1}{p}$). However $|-1|_p = 1$, which gives a contradiction. $\square$

Corollary 5.4.14 *The Factorial Conjecture $FC(R, n)$ holds for $n = 1$ and every commutative ring R contained in a $\mathbb{Q}$-algebra.*

Proof By Exercise 1 of Sect. 5.2 it suffices to prove $FC(\mathbb{C}, 1)$. So we need to show that $r(\ker \mathfrak{L}) = 0$, where $\mathfrak{L} : \mathbb{C}[t] \to \mathbb{C}$ is the $\mathbb{C}$-linear map defined by $\mathfrak{L}(t^n) = n!$ for all $n \in \mathbb{N}$. Since the $\mathbb{C}$-linear map L defined in Proposition 5.4.13 also satisfies $L(t^n) = n!$ for all $n \in \mathbb{N}$ we get $\mathfrak{L} = L$. Hence the result follows from Proposition 5.4.13. $\square$

Proposition 5.4.15 *Let $L : \mathbb{C}[t] \to \mathbb{C}$ be defined by $L(f(t)) = \int_{-\infty}^{\infty} f(t)e^{-t^2}\, dt$. Then $r(\ker L) = \{0\}$.*

Proof For a non-negative integer k we define $k!! = 0$ if k is odd and $k!! = (k-1)\cdot(k-3)\cdots 3\cdot 1$ if k is even. Then one verifies that $L(t^k) = \sqrt{\pi}(1/2)^{k/2}k!!$ for all $k \geq 0$. So if we replace L by $L_1 := \frac{1}{\sqrt{\pi}}L$, then $L_1(t^k) \in \mathbb{Q}$ for all $k \geq 0$ and $\ker L = \ker L_1$.

Now suppose that $r(\ker L) \neq 0$. Then $r(\ker L_1) \neq 0$. So there exist $0 \neq f \in \mathbb{C}[t]$ and $N \geq 1$ such that $L_1(f^m) = 0$ for all $m \geq N$. By Proposition 5.4.10 we may assume that $f \in K[t]$, where K is some number field. Also we may assume that $f = t^r + c_{r+1}t^{r+1} + \cdots + c_d t^d$, with $d > r$. By replacing f by f^2 we may additionally assume that r is even. For each prime number p we get

$$f^p = t^{rp} + \sum_{i=r+1}^{d} c_i^p t^{ip} + p\sum_{i=rp+1}^{dp-1} h_i(c)t^i,$$

where $h_i(c)$ belongs to the subring of K, generated by the c_i over $\mathbb{Z}$. Now choose p large enough such that $L_1(f^p) = 0$ and choose a p-adic absolute value $|.|_p$ such that $|c_i|_p \leq 1$ for all i. Then $|h_i(c)|_p \leq 1$ for all i. Applying L_1 to the equation above gives

$$0 = (\tfrac{1}{2})^{rp}(rp)!! + \sum_{i=r+1}^{d} c_i^p(\tfrac{1}{2})^{ip}(ip)!! + p\sum_{i=rp+1}^{dp-1} h_i(c)(\tfrac{1}{2})^i i!!.$$

Observe that if $i \geq r+1$, then $(ip)!!/(rp)!! = n_i p$ for some integer n_i and that $i!!/(rp)!! = m_i$ for some integer m_i if $i \geq rp + 1$. Now dividing the last equality by $(\tfrac{1}{2})^{rp}(rp)!!$ we obtain

$$0 = 1 + \sum_{i=r+1}^{d} c_i^p(\tfrac{1}{2})^{ip-rp}n_i p + p\sum_{i=rp+1}^{dp-1} h_i(c)(\tfrac{1}{2})^{i-rp}m_i.$$

Finally observe that each term of the two sums has a p-adic absolute value < 1, since $|p|_p < 1$, both $|c_i|_p, |h_i(c)|_p \leq 1$ for all i and $|2|_p = 1$. Since $|1|_p = 1$ this gives a contradiction. □

Exercises for Sect. 5.4

1. Let k be a field and $k[T]$ the univariate polynomial ring. Let $f \in k[T]$ with $f(0) = 1$ and denote by $g = f^{-1}$ the multiplicative inverse of f in $k[[T]]$. Write $g = \sum_{i=0}^{\infty} g_i T^i$, with $g_i \in k$. Show that if there exists an $r \geq 1$ such that $g_{mr} = 0$ for all $m \geq 1$, then $f = 1$ [The hypothesis on g implies that $g = 1 + a_1 T + \cdots + a_{r-1}T^{r-1}$, with $a_i \in k[[T^r]]$. View everything inside $k((T)) = \bigoplus_{i=0}^{r-1} k((T^r))T^i$ and deduce from

$g = 1/f \in k(T) = \bigoplus_{i=0}^{r-1} k(T^r)T^i$ that $a_i \in k(T^r)$ for all i. Let $0 \neq q \in k[T^r]$ be such that $qa_i \in k[T^r]$ for all i. Deduce from $q = (qg)f$ that $deg_T f = 0$].

2. Let $\mathbb{Z}[C] := \mathbb{Z}[C_1, \ldots, C_n]$ be the polynomial ring in the C_i over $\mathbb{Z}$. For each $t \geq 1$ define

$$h_t := \sum_{|i|=t} C_1^{i_1} \cdots C_n^{i_n},$$

where $|i| = i_1 + \cdots + i_n$. Let $e_1, \ldots, e_n$ be the elementary symmetric polynomials in the C_i.
   (i) Show that $S(T) := \prod_{i=1}^n (1 - C_i T) = 1 - e_1 T + e_2 T^2 + \cdots + (-1)^n e_n T^n$.
   (ii) Show that $1/S(T) = 1 + \sum_{t=1}^\infty h_t T^t \in \mathbb{Z}[C][[T]]$.

3. Let k be a field and $f = g^r$, where $r \geq 1$ and $g = c_1 X_1 + \cdots + c_n X_n$ with all c_i in k. Show that f satisfies the Factorial Conjecture, i.e., if $\mathfrak{L}(f^m) = 0$ for all large m, then $f = 0$ [Use Exercise 1 and 2].

4. This exercise is a generalization of Exercise 1. Let A be a commutative ring with nilradical $\mathfrak{n}$. By $A[T]$, respectively $A[[T]]$, we denote the polynomial ring, respectively the power series ring, in the variables $T_1, \ldots, T_n$ over A. If $r = (r_1, \ldots, r_n)$, where each r_i is a positive integer, we can write every $g \in A[[T]]$ uniquely on the form

$$g = g_0 + \sum_{i < r, i \neq 0} g_i T^i, \quad \text{with } g_i \in A[[T^r]],$$

where the elements of $A[[T^r]]$ are (infinite) sums of monomials of the form $aT_1^{i_1 r_1} \cdots T_n^{i_n r_n}$. We say that g has **nilpotent r-gaps** if for each monomial aT^i appearing in g_0, with $i \neq 0$ we have that $a \in \mathfrak{n}$. Show that f is a unit in $A[T]$ if and only if g has nilpotent r-gaps for some r as above.

Notes

The Classical Orthogonal Polynomial Conjecture is Conjecture 3.5 in [138]. The Gaussian Moments Conjecture and the related results Theorem 5.4.3 and the propositions 5.4.4 and 5.4.5 come from the paper [31]. Also Example 5.4.7 is taken from [31]. Finally, the Strong Integral Conjecture and the related Theorem 5.4.9 can be found in [125].

5.5 The Duistermaat–van der Kallen Theorem

In Sect. 5.2 we formulated the Mathieu Conjecture, which formed one of the main motivations for Zhao to introduce his Mathieu–Zhao spaces. However until now only one case has been proved, namely the case that the group G is abelian. In their landmark paper [40] Duistermaat and van der Kallen proved this case and their proof is highly nontrivial. In more algebraic terms their result can be formulated as follows:

let $\mathbb{C}[x, x^{-1}] := \mathbb{C}[x_1, \ldots, x_n, x_1^{-1}, \ldots, x_n^{-1}]$ be the ring of Laurent polynomials over $\mathbb{C}$ and if $f \in \mathbb{C}[x, x^{-1}]$ we denote by f_0 the coefficient of the constant term of f.

Theorem 5.5.1 (Duistermaat–van der Kallen) *The set of Laurent polynomials $f \in \mathbb{C}[x, x^{-1}]$ such that $f_0 = 0$ is an MZ-space of $\mathbb{C}[x, x^{-1}]$.*

To prove this result Duistermaat and van der Kallen proved a more stronger result, namely that the Newton polytope of a Laurent polynomial f which has the property that $(f^m)_0 = 0$ for all $m \geq 1$ does not contain 0. Using this result we will deduce an equivalent formulation, the so-called **Density theorem**, which turns out to be very useful for several applications.

Polytopes

First we recall some well-known properties of Newton polytopes. We refer the reader to [12] for a more detailed treatment of these objects.

Let $A = \{a_1, \ldots, a_r\} \subseteq \mathbb{R}^n$. The **polytope of** A, denoted $Poly(A)$, is the convex hull of A in $\mathbb{R}^n$, i.e., the set of all $\mathbb{R}$-linear combinations $c_1 a_1 + \cdots + c_r a_r$, with $c_i \geq 0$ for all i and $c_1 + \cdots + c_r = 1$. We can add polytopes and multiply them by scalars from $\mathbb{R}$ to obtain new polytopes. More precisely, if $c \in \mathbb{R}$ and P and Q are polytopes in $\mathbb{R}^n$, one can define their **Minkowski sum**

$$P + Q := \{p + q \mid p \in P, q \in Q\}$$

and

$$cP := \{cp \mid p \in P\}.$$

Both $P + Q$ and cP are again polytopes. Furthermore, if $m \geq 1$ is an integer, mP equals the m-fold sum $P + P + \cdots + P$. If all a_i belong to $\mathbb{Q}^n$ the corresponding polytope is called a **rational polytope**. The sum of rational polytopes is again a rational polytope and the same holds for the intersection if it is non-empty. Furthermore, if P and Q are rational polytopes and u is a rational point of $P - Q$, then there exist rational points $p \in P$ and $q \in Q$ such that $u = p - q$.

In most of our applications the polytopes arise from Laurent polynomials: let $0 \neq f \in \mathbb{C}[x, x^{-1}]$. Write $f = \sum_{a \in \mathbb{Z}^n} c_a x^a$, with $c_a \in \mathbb{C}$. The **support of** f, denoted $Supp(f)$, is the set of all $a \in \mathbb{Z}^n$ such that $c_a \neq 0$. The **Newton polytope of** f, denoted $Poly(f)$, is the polytope of $Supp(f)$. If both f and g are nonzero Laurent polynomials, then $Poly(fg) = Poly(f) + Poly(g)$. From this one deduces that $Poly(x^b f^m) = b + m Poly(f)$ for all $m \geq 1$ and $b \in \mathbb{Z}^n$.

The Density Theorem

Now we are able to formulate the main result of this section, namely Theorem 5.5.2. A proof of this result in dimension one is given at the end of this section. For the general case we refer to [40].

Theorem 5.5.2 (Duistermaat–van der Kallen) *Let $f \in \mathbb{C}[x, x^{-1}]$. Then $(f^m)_0 = 0$ for all $m \geq 1$ if and only if $0 \notin Poly(f)$.*

Proof of Theorem 5.5.1 Let $f \in \mathbb{C}[x, x^{-1}]$ be such that $(f^m)_0 = 0$ for all $m \geq 1$. It suffices to show that for each $b \in \mathbb{Z}^n$ we have that $(x^b f^m)_0 = 0$ for all large m. By Theorem 5.5.2 $0 \notin Poly(f)$. Denote by δ the distance function in $\mathbb{R}^n$. Since $Poly(f)$ is compact $d := \delta(0, Poly(f)) > 0$. So $\delta(0, Poly(f^m)) = \delta(0, m\,Poly(f)) = md$. Denote by $|b|$ the length of b. Then for all large enough m we get $\delta(0, Poly(x^b f^m)) = \delta(0, b + m\,Poly(f)) \geq md - |b| > 0$. In particular $0 \notin Supp(x^b f^m)$ for all large m, i.e., $(x^b f^m)_0 = 0$ for all large m, which completes the proof.   □

Another consequence of Theorem 5.5.2 is:

Corollary 5.5.3 *If $0 \in Poly(f)$, then $(f^m)_0 \neq 0$ for infinitely many $m \geq 1$.*

Proof   Assume the contrary. Then for some $N \in \mathbb{N}$ we have $(f^m)_0 = 0$ for all $m \geq N$. Put $g := f^N$. Then $(g^m)_0 = 0$ for all $m \geq 1$. So by Theorem 5.5.2 $0 \notin Poly(g) = N\,Poly(f)$. So $0 \notin Poly(f)$, a contradiction.   □

Theorem 5.5.4 (Density Theorem) *Let $P \in \mathbb{C}[x, x^{-1}]$ and $u \in \mathbb{Q}^n \cap Poly(P)$. Then $mu \in Supp(P^m)$ for infinitely many $m \geq 1$.*

Proof   Let $N \in \mathbb{N}$, $N \geq 1$, be such that $Nu \in \mathbb{Z}^n$. Then $Nu \in N\,Poly(P) = Poly(P^N)$. Put $f := x^{-Nu} P^N$. Then $0 \in Poly(f)$. By Corollary 5.5.3 we get $0 \in Supp(f^m) = -mNu + Supp(P^{Nm})$ for infinitely many $m \geq 1$, whence $(mN)u \in Supp(P^{mN})$ for infinitely many m.   □

Remark 5.5.5 The above arguments show that Theorem 5.5.2 implies the Density theorem. Conversely Theorem 5.5.2 follows immediately from the Density theorem by taking $u = 0$. So both theorems are equivalent.

Applications of the Density Theorem

To demonstrate the power of the Density theorem we give two applications. First we will show that the Generalized Vanishing Conjecture is true for the monomial operator $\partial^a :=$

$\partial_1^{a_1} \cdots \partial_n^{a_n}$. To do so we make some easy observations. Let $a \in \mathbb{N}^n$ and $Q \in \mathbb{C}[x]$. Then one readily verifies that

$$\partial^a Q = 0 \text{ if and only if } Supp(x^{-a} Q) \cap \mathbb{R}_{\geq 0}^n = \emptyset.$$

So if $\Lambda = \partial^a$, $b \in \mathbb{N}^n$ and $P \in \mathbb{C}[x]$, then

$$\Lambda^m(x^b P^m) = 0 \text{ if and only if } Supp(x^b f^m) \cap \mathbb{R}_{\geq 0}^n = \emptyset,$$

where $f := x^{-a} P \in \mathbb{C}[x, x^{-1}]$. Taking $b = 0$ we get

$$\Lambda^m(P^m) = 0 \text{ if and only if } Supp(f^m) \cap \mathbb{R}_{\geq 0}^n = \emptyset.$$

The second observation is the following: let $D \subseteq \mathbb{R}^n$ be compact and $G \subseteq \mathbb{R}^n$ be closed. If $D \cap G = \emptyset$, there exist $d_0 \in D$ and $g_0 \in G$ such that $0 < \delta(d_0, g_0) \leq \delta(d, g)$, for all $d \in D$ and $g \in G$. This minimal distance is called the **distance between D and G** and will be denoted by $\delta(D, G)$.

After these preparations we are able to prove:

Theorem 5.5.6 *Let $\Lambda = \partial^a$ and $P \in \mathbb{C}[x]$ be such that $\Lambda^m P^m = 0$ for all $m \geq 1$. Then for each $Q \in \mathbb{C}[x]$ there exists $N \geq 1$ such that $\Lambda^m(QP^m) = 0$ for all $m \geq N$.*

Proof   We may assume $Q = x^b$, with $b \in \mathbb{N}^n$. Let $f := x^{-a} P$. By the observations above $Supp(f^m) \cap \mathbb{R}_{\geq 0}^n = \emptyset$ for all $m \geq 1$. We claim: $Poly(f) \cap \mathbb{R}_{\geq 0}^n = \emptyset$. Namely, if this is not the case there exists $u \in Poly(f) \cap \mathbb{Q}_{\geq 0}^n$. So by Theorem 5.5.4 $mu \in Supp(f^m) \cap \mathbb{R}_{\geq 0}^n$ for infinitely many $m \geq 1$, a contradiction. So $d := \delta(Poly(f), \mathbb{R}_{\geq 0}^n) > 0$. Hence

$$\delta\left(Poly(x^b f^m), \mathbb{R}_{\geq 0}^n\right) = \delta\left(b + m Poly(f), \mathbb{R}_{\geq 0}^n\right) \geq md - |b| > 0$$

if m is large enough. In particular $Supp(x^b f^m) \cap \mathbb{R}_{\geq 0}^n = \emptyset$ for all m large enough, which, as observed above, completes the proof. $\square$

The second application of the Density theorem shows that the image of the derivation $D := \sum_i a_i x_i \partial_{x_i}$ ($a_i \in \mathbb{C}$) is an MZ-space of $\mathbb{C}[x]$. Before we prove this result we make some observations. First of all we may assume that not all a_i are zero (otherwise $Im D = 0$ and we are done). Let $b \in \mathbb{N}^n$. Then $Dx^b = \langle a, b \rangle x^b$, where $\langle , \rangle$ denotes the usual inner product on $\mathbb{R}^n$. So $x^b \in Im D$ if and only if $\langle a, b \rangle \neq 0$. Define S to be the free $\mathbb{Z}$-module consisting of all $b \in \mathbb{Z}^n$ satisfying $\langle a, b \rangle = 0$. Let $u_1, \ldots, u_r$ be free $\mathbb{Z}$-generators of this module. Then the u_i form an $\mathbb{R}$-basis of the $\mathbb{R}$-vector space V generated by S and its dimension r is smaller than n (not all a_i are zero). Furthermore $x^b \in Im D$ if and only if $b \notin S$.

Claim: $b \notin S$ if and only if $b \notin V$. Namely, if $b \in V$, then b is an $\mathbb{R}$-linear combination of the u_i. Since each u_i belongs to S we get $\langle a, u_i \rangle = 0$ for all i. Hence $\langle a, b \rangle = 0$, i.e., $b \in S$. Summarizing: $x^b \in ImD$ if and only if $b \notin V$. Consequently, for any $0 \neq h \in \mathbb{C}[x]$ we get

$$h \in ImD \text{ if and only if } Supp(h) \cap V = \emptyset.$$

Theorem 5.5.7 *Let $D := \sum_i a_i x_i \partial_{x_i}$, with $a_i \in \mathbb{C}$. Then ImD is an MZ-space of $\mathbb{C}[x]$.*

Proof Let $0 \neq f \in \mathbb{C}[x]$ be such that $f^m \in ImD$ for all $m \geq 1$. Then, by the observations above, $Supp(f^m) \cap V = \emptyset$ for all $m \geq 1$. We claim: $Poly(f) \cap V = \emptyset$. Assume otherwise. It then follows from Lemma 5.5.8 below that there exists $u \in \mathbb{Q}^n$ such that $u \in Poly(f) \cap V$. Then Theorem 5.5.4 implies that $mu \in Supp(f^m) \cap V$ for infinitely many m, a contradiction. So $Poly(f) \cap V = \emptyset$. Hence, as in the proof of Theorem 5.5.6, we deduce that $d := \delta(Poly(f), V) > 0$. Furthermore, if $b \in \mathbb{N}^n$ we deduce that $\delta(Poly(x^b f^m), V) \geq md - |b| > 0$ for all m large enough. In particular $Supp(x^b f^m) \cap V = \emptyset$, which implies that $x^b f^m \in ImD$ for all m large enough. This completes the proof. $\qquad\square$

Lemma 5.5.8 *Notations as above. If $\Sigma := Poly(f) \cap V \neq \emptyset$, then Σ is a rational polytope (and hence contains an element of $\mathbb{Q}^n$).*

Proof Extend $u_1, \ldots, u_r$ to an $\mathbb{R}$-basis $u_1, \ldots, u_n$ of $\mathbb{R}^n$ and let $T : \mathbb{R}^n \rightarrow \mathbb{R}^n$ be the $\mathbb{R}$-linear map defined by $T(u_i) = e_i$ for all i, where the e_i form the standard basis of $\mathbb{R}^n$. Since $Poly(f)$ is compact, the same holds for $T(\Sigma)$. So there exists $N \geq 1$ such that $T(\Sigma) \subseteq B_N(0)$, the sphere with radius N around 0. Hence if $a_1 u_1 + \cdots + a_r u_r \in \Sigma$, then $T(a_1 u_1 + \cdots + a_r u_r) = (a_1, \ldots, a_r, 0, \ldots, 0) \in B_N(0)$. So $|a_i| \leq N$ for all $1 \leq i \leq r$, whence $\Sigma = Poly(f) \cap (\mathbb{R}u_1 + \cdots + \mathbb{R}u_r) \subseteq Poly(f) \cap (\sum_{i=1}^{r}[-N, N]u_i) \subseteq \Sigma$. So $\Sigma = Poly(f) \cap (\sum_{i=1}^{r}[-N, N]u_i)$, an intersection of rational polytopes, which implies that Σ is a rational polytope. $\qquad\square$

Generalizing the Duistermaat–van der Kallen Theorems

Let R be a commutative ring. A **Laurent polynomial** f in the variables $x = (x_1, \ldots, x_n)$ over R is a finite R-linear combination of the monomials $x^m := x_1^{m_1} \cdots x_n^{m_n}$, where $m = (m_1, \ldots, m_n) \in \mathbb{Z}^n$. The coefficient of x^m in f will be denoted bij f_m. The set of Laurent polynomials in x over R is a ring, denoted by $R[x, x^{-1}]$.

The aim of this section is to investigate the question: can we replace $\mathbb{C}$ in Theorem 5.5.1 by an arbitrary commutative ring R and obtain a similar result? The first result, which was obtained in [140], shows that such a result does not hold in any characteristic. More precisely:

Theorem 5.5.9 *Let k be a field of characteristic $p > 0$ and $f := t^{-1} + t^{p-1} \in k[t, t^{-1}]$, the univariate Laurent polynomial ring. Then $(f^m)_0 = 0$ for all $m \geq 1$, however $(t^{-1} f^m)_0 = (-1)^m$ for all m of the form $p^n - 1$ with $n \geq 1$. So the set of f in $k[t, t^{-1}]$ with $f_0 = 0$ is not an MZ-space of $k[t, t^{-1}]$.*

Proof

(i) Let $m \geq 1$. Then $(f^m)_0$ is the sum of all $\binom{m}{i}$, with $0 \leq i \leq m$ such that $-(m - i) + i(p - 1) = 0$. The only possible such an i is m/p, in case p divides m. It then follows from Lemma 5.5.10(i) below that $(f^m)_0 = 0$.

(ii) Let $n \geq 1$ and $m = p^n - 1$. Then the coefficient of t in f^m equals $\binom{m}{i}$, where $i = p^{n-1}$. Hence $(t^{-1} f^m)_0 \neq 0$ by Lemma 5.5.10(ii).   □

Lemma 5.5.10

(i) Let $i \geq 1$. Then $\binom{ip}{i}$ is zero in $\mathbb{F}_p$.

(ii) Let $n \geq 1$ and $m = p^n - 1$. Then $\binom{m}{i} = (-1)^i$ in $\mathbb{F}_p$ for all $0 \leq i \leq m$.

Proof

(i) Write $i = p^r n$ with $r \geq 0$ such that p^{r+1} does not divide i. The coefficient of t^i in $(t + 1)^{ip} \in \mathbb{F}_p[t]$ equals $\binom{ip}{i}$. Furthermore

$$(t + 1)^{ip} = (t + 1)^{p^{r+1} n} = (t^{p^{r+1}} + 1)^n.$$

So if $\binom{ip}{i}$ is nonzero in $\mathbb{F}_p$ the monomial t^i appears in $(t + 1)^{ip}$. Hence $i = p^{r+1} j$ for some $1 \leq j \leq n$. This contradicts the fact that p^{r+1} does not divide i.

(ii) First observe that $(1 - t)^m = \sum_{i=0}^m (-1)^i \binom{m}{i} t^i$. Furthermore in $\mathbb{F}_p(t)$ we have

$$(1 - t)^m = \frac{(1 - t)^{p^n}}{1 - t} = \frac{1 - t^{p^n}}{1 - t} = \sum_{i=0}^{p^n - 1} t^i.$$

Comparing the coefficients of t^i in both equations gives the result.   □

So from now on we assume: R is contained in a $\mathbb{Q}$-algebra S. Furthermore, if no confusion is possible, we denote the nilradical $\mathfrak{n}(R)$ of R by $\mathfrak{n}$. With these conventions we are able to generalize the Duistermaat–van der Kallen theorem as follows:

Theorem 5.5.11 *Let M be the set of f in $R[x, x^{-1}]$ such that $f_0 \in \mathfrak{n}$. Then M is an MZ-space of $R[x, x^{-1}]$.*

Proof   Let $f \in R[x, x^{-1}]$ such that $(f^m)_0 \in \mathfrak{n}$ for all $m \geq 1$. We want to show that for each $b \in \mathbb{Z}^n$ also $(x^b f^m)_0 \in \mathfrak{n}$ for all large m.

(i) First we prove the case that R is a domain of characteristic zero. Let R_0 be the $\mathbb{Z}$-subalgebra of R generated by the coefficients of f. Then by Lefschetz principle its quotient field can be embedded into $\mathbb{C}$. So we can view f inside $\mathbb{C}[x, x^{-1}]$. Then the Duistermaat–van der Kallen theorem implies that $(x^b f^m)_0 = 0$ for all large m. So M is an MZ-space of $R[x, x^{-1}]$.

(ii) Now the general case. Let R_0 be the $\mathbb{Q}$-subalgebra of S generated by the coefficients of f. Then R_0 is a Noetherian $\mathbb{Q}$-algebra, $f \in R_0[x, x^{-1}]$ and $(f^m)_0 \in \mathfrak{n}(R_0)$ for all $m \geq 1$. So, replacing R by R_0, we may assume that R is a Noetherian $\mathbb{Q}$-algebra. This implies that $\mathfrak{n} = \mathfrak{p}_1 \cap \cdots \cap \mathfrak{p}_s$, for some finite set of prime ideals $\mathfrak{p}_i$ of R.

Now let $b \in \mathbb{Z}^n$. It suffices to show that for each $1 \leq i \leq s$ there exists $N_i \geq 1$ such that $(x^b f^m)_0 \in \mathfrak{p}_i$ for all $m \geq N_i$ (for then take N to be the maximum of all N_i and use that $\mathfrak{n} = \mathfrak{p}_1 \cap \cdots \cap \mathfrak{p}_s$). So let $1 \leq i \leq s$. Since R is a $\mathbb{Q}$-algebra, $R/\mathfrak{p}_i$ is a domain of characteristic zero. Furthermore $\overline{f} \in R/\mathfrak{p}_i[x, x^{-1}]$ satisfies $(\overline{f}^m)_0 = 0$ for all $m \geq 1$. It then follows from i) that there exists $N_i \geq 1$ such that $(x^b \overline{f}^m)_0 = 0$ for all $m \geq N_i$. Hence $(x^b f^m)_0 \in \mathfrak{p}_i$ for all $m \geq N_i$, which concludes the proof. $\square$

Next we investigate the stronger Theorem 5.5.2. Using Lefschetz principle (as in the proof of Theorem 5.5.11 above) one immediately deduces from Theorem 5.5.2:

Theorem 5.5.12   *Let R be a domain of characteristic zero and $f \in R[x, x^{-1}]$ such that $(f^m)_0 = 0$ for all $m \geq 1$. Then $0 \notin Poly(f)$.*

However, if R is not a domain the statement of Theorem 5.5.2 fails:

Example 5.5.13   Let $a, b \in R \backslash \{0\}$ with $ab = 0$ and $f := at^{-1} + bt \in R[t, t^{-1}]$. Then $f^m = a^m t^{-m} + bt^m$, so $(f^m)_0 = 0$ for all $m \geq 1$.

To conclude this section we will give a proof of Theorem 5.5.12 in dimension one. So $R[x] = R[x_1]$ and R is a domain of characteristic zero. By Lefschetz principle we may assume that $R = \mathbb{C}$. So let $f = \sum_i f_i x^i \in \mathbb{C}[x, x^{-1}]$. By $l_0 : \mathbb{C}[x, x^{-1}] \to \mathbb{C}$ we denote the $\mathbb{C}$-linear map defined by $l_0(f) = f_0$.

One readily verifies that the statement of Theorem 5.5.12 (with $R = \mathbb{C}$) is equivalent to:

Theorem 5.5.14   *If $l_0(f^m) = 0$ for all $m \geq 1$, then $f \in \mathbb{C}[x]$ or $f \in \mathbb{C}[x^{-1}]$.*

Proof

(i) Assume $f \notin \mathbb{C}[x]$ and $f \notin \mathbb{C}[x^{-1}]$. Then $f = \alpha x^{-s} + \cdots + \beta x^r$ with $r, s \geq 1$ and $\alpha, \beta \in \mathbb{C}^*$. Define the generating function $W(z) := \sum_{m \geq 0} l_0(f^m) z^m$. We will show that $W(z) \neq 1$, which contradicts the hypothesis that $l_0(f^m) = 0$ for all $m \geq 1$. To show this we want to write $W(z) = l_0(1/1 - zf)$. In order to do so we first extend the map l_0.

(ii) Put $U(x) := x^s(1 - zf(x)) \in \mathbb{C}(z)[x] \subseteq \mathbb{C}((z))[x]$ and $n := r + s$. Then (Exercise 3) there exists $p \geq 1$ such that

$$U(x) = (-\beta z)(x - a_1) \cdots (x - a_n), \text{ with all } a_i \in \mathbb{C}((z^{\frac{1}{p}})).$$

Put $k := \mathbb{C}((z^{\frac{1}{p}}))$. On k we have the valuation v defined by $v(z^{\frac{1}{p}}) = \frac{1}{p}$ and k is complete with respect to this valuation (Exercise 6). From Exercises 5 and 6 it follows that $k[[x, x^{-1}]]$ is a ring. Define

$$l_0 : k[[x, x^{-1}]] \to k \text{ by } l_0\left(\sum_{n=-\infty}^{\infty} c_n x^n\right) = c_0.$$

By Exercise 6 $\mathbb{C}[x, x^{-1}][[z]]$ is a subring of $k[[x, x^{-1}]]$. Hence $\sum_{m \geq 0} f^m z^m \in k[[x, x^{-1}]]$. So $1 - zf$ is invertible in $k[[x, x^{-1}]]$ and hence so is $U(x)$ and each factor $x - a_i$. We denote the inverse of $1 - zf$ by $1/1 - zf$ and obtain

$$W(z) = l_0\left(\frac{1}{1 - zf}\right) = l_0\left(\frac{x^s}{U(x)}\right).$$

(iii) To compute the last expression we will use a partial fraction decomposition of $x^s/U(x)$. Therefore observe that since $(-1)^n a_1 \cdots a_n = \alpha/\beta$, all $a_i \neq 0$. From $U(a_i) = 0$ and $a_i \neq 0$ we get $f(a_i) = 1/z$. So $f'(a_i)a_i' = -z^{-2}$. Since $U'(x) = sx^{s-1}(1 - zf(x)) + x^s(-z)f'(x)$ we get $U'(a_i) = (-z)a_i^s f'(a_i) \neq 0$. So all a_i are different. Since $s < s + r = n$ it follows from Exercise 4 that we get a partial fraction decomposition of the form

$$\frac{x^s}{U(x)} = \sum_{i=1}^{n} \frac{A_i}{x - a_i},$$

where $A_i = \frac{a_i^s}{U'(a_i)} = -\frac{1}{zf'(a_i)}$ (use that $U'(a_i) = -a_i^s zf'(a_i)$). So

$$\frac{1}{1 - zf} = \frac{x^s}{U(x)} = \sum_{i=1}^{n} -\frac{1}{zf'(a_i)(x - a_i)}. \tag{5.5.1}$$

(iv) Next we compute the inverse of $x - a_i$ in $k[[x, x^{-1}]]$. Observe that $f(a_i) = 1/z$ implies that $v(a_i) \neq 0$. So either $v(a_i) > 0$ or $v(a_i) < 0$.
If $v(a_i) > 0$, then

$$(x - a_i)^{-1} = x^{-1}(1 - a_i x^{-1})^{-1} = x^{-1} \sum_{m=0}^{\infty} a_i^m x^{-m} \in k[[x, x^{-1}]].$$

If $v(a_i) < 0$, then

$$(x - a_i)^{-1} = -a_i^{-1}(1 - a_i^{-1}x)^{-1} = -a_i^{-1} \sum_{m=0}^{\infty} (a_i^{-1}x)^m \in k[[x, x^{-1}]].$$

So, if S denotes the set of i such that $v(a_i) < 0$, then we get from (5.5.1) that

$$W(z) = l_0\left(\frac{1}{1 - zf}\right) = \sum_{i \in S} \frac{1}{za_i f'(a_i)}.$$

Since $f'(a_i)a_i' = -z^{-2}$ we get

$$W(z) = \sum_{i \in S} \frac{1}{za_i f'(a_i)} = -z \sum_{i \in S} \frac{a_i'}{a_i}.$$

It remains to show that

$$\sum_{i \in S} \frac{a_i'}{a_i} \neq -\frac{1}{z}. \tag{5.5.2}$$

(v) To show this inequality we study the a_i at infinity. More precisely, write $a := a_i$ and put $t := 1/z$. Then $f(a) = t$. Since $\mathbb{C}(z) = \mathbb{C}(t)$ it follows that a is algebraic over $\mathbb{C}(t)$ and hence over $\mathbb{C}((t))$. Therefore we can view a inside $\mathbb{C}((t^{\frac{1}{p}}))$ for some $p \geq 1$ (Exercise 3). Since $a \neq 0$ we get $a = \sum_{j=m}^{\infty} c_j t^{j/p}$, with $c_m \in \mathbb{C}^*$. Let w denote the valuation on $\mathbb{C}((t^{\frac{1}{p}}))$ with $w(t^{\frac{1}{p}}) = 1/p$. Then $w(a) = m/p$. Now we claim that $w(a) = 0$. Namely, if $w(a) > 0$, then $w(f(a)) = w(\alpha c_m^{-s} t^{-ms/p}) = -ms/p = -sw(a) < 0$, since $s \geq 1$. This is a contradiction since $f(a) = t$ and $w(t) = 1$. Similarly, if $w(a) < 0$ then $w(f(a)) = w(c_m^r t^{mr/p}) = mr/p = w(a)r < 0$, since $r \geq 1$. So $w(a) = 0$. Consequently $a = \sum_{j=0}^{\infty} c_j t^{j/p}$, with $c_0 \in \mathbb{C}^*$.

(vi) Finally, let us return to the expression in (5.5.2). The right-hand side equals $-t$ and $w(-t) = 1$. Therefore, to prove the inequality in (5.5.2), it suffices to show that

$w(a'/a) > 1$, for each a appearing in the sum. Observe that

$$\frac{a'(z)}{a(z)} = -\frac{\frac{da}{dt}}{a} t^2.$$

Since $f(a) = t$ we get $a \notin \mathbb{C}$. So there exists $j > 0$ such that $c_j \neq 0$. Choose j minimal with this property. Then $a = c_0 + c_j t^{j/p} + R$, with $w(R) > j/p$. Hence $w(\frac{da}{dt}) = \frac{j}{p} - 1$. Since $w(a) = 0$, as shown above, it follows that

$$w\left(\frac{a'(z)}{a(z)}\right) = \frac{j}{p} - 1 + 2 > 1$$

which concludes the proof.   □

Exercises for Sect. 5.5

1. Let R be a domain of characteristic zero and $R[t, t^{-1}]$ the ring of Laurent polynomials in one variable over R. Denote by M the set of $f = \sum_i f_i t^i \in R[t, t^{-1}]$ such that $f_0 = 0$. Show that $r(M) = t R[t] \cup t^{-1} R[t^{-1}]$ [Use Theorem 5.5.12].
2. Let $z = x + iy$ and $\bar{z} = x - iy$ in $\mathbb{C}[x, y] = \mathbb{C}[z, \bar{z}]$. Let D be the unit disk around the origin in the plane and define

$$M_D = \{f \in \mathbb{C}[x, y] \mid \int_D f(x, y)\, dxdy = 0\}.$$

(i) Let $P = P(z, \bar{z})$ be homogeneous of degree d. Show that for all $m \geq 1$

$$\int_D P(z, \bar{z})^m \, dxdy = \frac{1}{dm + 2} \int_{t=0}^{2\pi} (P(u, u^{-1})^m)_{|u=e^{it}}\, dt$$

$$= \frac{2\pi}{dm + 2} P(u, u^{-1})^m{}_0.$$

(ii) Deduce from (i) and Exercise 1 above that if $P \in r(M_D)$ is homogeneous, then $P \in sr(M_D)$.
3. (Newton–Puiseux theorem). Let k be an algebraically closed field of characteristic zero and $f(x) = \sum_{i=0}^n a_i(t) x^i \in k((t))[x]$, with $n := \deg_x f(x) \geq 1$. We will show that there exists $p \geq 1$ such that $f(x)$ splits completely in linear factors over $k((t^{1/p}))$. By v we denote the standard valuation on $k((t))$ and use the same notation for its extension to larger fields such as $k((t^{1/p}))$.
   (i) Show that we may assume that $f(x)$ is monic and $v(a_i(t)) \geq 0$ for all i.
   (ii) Put $g(x) := f(x - a_{n-1}(t)/n)$. If $g(x) = x^n$ we are done. So assume $g(x) := \sum_{i=0}^n b_i(t) x^i \neq 0$. Show that $b_{n-1}(t) = 0$ and $v(b_i(t)) \geq 0$ for all i.

(iii) Put $q := min_{1 \leq i \leq n} \frac{v(b_{n-i}(t))}{i}$. Write $q = r/s$, with $r, s \geq 1$. Show that $t^{-nq} g(x) = h(y)$, where $y := t^{-q} x$ and $h(y) = \sum_{i=0}^{n} c_i(t) y^i$, with $c_i(t) \in k[[t^{1/s}]]$, $v(c_i(t)) \geq 0$ and $v(c_i(t)) = 0$ for at least one i.

(iv) Let $c_i(0)$ be the constant term of $c_i(t)$. Show that $\sum_{i=0}^{n} c_i(0) y^i$ has at least two different zeros in k (remember that $b_{n-1}(t) = 0$!) Deduce from Hensel's lemma that there exist $h_1(y), h_2(y) \in k[[t^{1/s}]][y]$ with $h(y) = h_1(y) h_2(y)$ and $deg_y h_1(y), h_2(y) < n$. Then use induction on n to complete the proof of the Newton–Puiseux theorem.

4. (Partial fractions decomposition). Let k be a field, $a_1, \ldots, a_n \in k$ all different and $\alpha \in k^*$. Put $U(x) = \alpha(x - a_1) \cdots (x - a_n)$.

  (i) Show that $U'(a_i) \neq 0$ for all i.

  (ii) Let $V(x) \in k[x]$ with $deg_x V(x) < n$. Show that there exist $A_1, \ldots, A_n \in k$ such that

$$\frac{V(x)}{U(x)} = \sum_i \frac{A_i}{x - a_i}.$$

  (iii) Deduce that $A_i = \frac{V(a_i)}{U'(a_i)}$ for all i.

5. Let k be a field and v a valuation on k, such that k is complete with respect to v. Define $k[[x, x^{-1}]]$ to be the set of formal series $\sum_{n=-\infty}^{\infty} c_n x^n$ such that $lim_{|n| \to \infty} c_n = 0$.

  (i) Let $a = \sum_{m=-\infty}^{\infty} a_m x^m$ and $b = \sum_{m=-\infty}^{\infty} b_m x^m$ in $k[[x, x^{-1}]]$. For $n \in \mathbb{Z}$ and every $N \geq 1$ define

$$c_{n,N} := \sum_{m=-N}^{N} a_m b_{n-m}.$$

Show that $c_{n,1}, c_{n,2}, c_{n,3}, \ldots$ is a Cauchy sequence in k. Since k is complete we can define $c_n := lim_{N \to \infty} c_{n,N}$, which we also denote by $\sum_{m=-\infty}^{\infty} a_m b_{n-m}$.

  (ii) Show that $lim \, c_n = 0$ if $|n| \to \infty$.

  (iii) Deduce that $k[[x, x^{-1}]]$ with the usual addition and multiplication defined by $ab := \sum_{n=-\infty}^{\infty} c_n x^n$, where $c_n = \sum_{m=-\infty}^{\infty} a_m b_{n-m}$, is a ring.

6. Let $k = \mathbb{C}((z^{\frac{1}{p}}))$ and v the valuation on k such that $v(z^{\frac{1}{p}}) = 1/p$.

  (i) Show that k is complete with respect to v.

  (ii) Let $k[[x, x^{-1}]]$ be as in Exercise 5. Show that $\mathbb{C}[x, x^{-1}][[z]]$ is a subring of $k[[x, x^{-1}]]$.

Notes

The density theorem was obtained in a somewhat weaker form as Theorem 4.6 in [128]. Theorem 5.5.4 is a generalized version of Corollary 4.8 in [128], where the case that f is homogeneous is proved. Theorem 5.5.6 is Corollary 5.3 in [128]. Theorem 5.5.7 is Lemma

3.4 in [126]. The proof of Theorem 5.5.14 is due to Paul Monsky. I am grateful to Lennaert Stronks for explaining this proof to me.

5.6   The Generalized Vanishing Conjecture

In the first section we introduced the Generalized Vanishing Conjecture for polynomial rings over a field of characteristic zero and showed how it implies the Jacobian Conjecture. More generally we have:

Generalized Vanishing Conjecture (GVC(R,n)) *Let R be a commutative ring contained in a $\mathbb{Q}$-algebra and Λ a differential operator with constant coefficients, i.e., $\Lambda \in D := R[\partial_1, \ldots, \partial_n]$. If $f \in R[x]$ is such that $\Lambda^m f^m = 0$ for all $m \geq 1$, then for each $g \in R[x]$ also $\Lambda^m(g f^m) = 0$ for all large m.*

Using Lefschetz principle one easily deduces

Proposition 5.6.1 *If $GVC(\mathbb{C}, n)$ is true, then $GVC(R, n)$ is true for every ring R that is contained in a commutative $\mathbb{Q}$-algebra.*

Therefore we restrict the study of this conjecture to the case that $R = \mathbb{C}$.

The main aim of this section is to give two equivalent formulations of the Generalized Vanishing Conjecture and to prove it in dimension two for *arbitrary homogeneous operators*.

We start by showing that the g's in the formulation of the Generalized Vanishing Conjecture can be replaced by powers of f. More precisely:

Theorem 5.6.2 (de Bondt) *The Generalized Vanishing Conjecture is equivalent to the following statement: if $f \in \mathbb{C}[x]$ is such that $\Lambda^m f^m = 0$ for all $m \geq 1$, then for each $d \geq 1$ $\Lambda^m(f^{d+m}) = 0$ for all large m.*

To prove this theorem we start making some observations concerning the Weyl algebra $A_n(\mathbb{C})$, the $\mathbb{C}$-algebra generated by $x_1, \ldots, x_n, \partial_1, \ldots, \partial_n$ which satisfy the relations

$$[x_i, x_j] = [\partial_i, \partial_j] = 0, [\partial_i, x_j] = \delta_{ij}, \text{ for all } i, j.$$

From these relations we obtain that if g is any polynomial in $\mathbb{C}[x]$ then $\partial_i g = g \partial_i + g_1$, where g_1 is a polynomial of degree $<\deg g$. It follows that for all $\alpha \in \mathbb{N}^n$ the operator $\partial^\alpha g - g \partial^\alpha$ belongs to $g_1 D$, the $\mathbb{C}$-vector space generated by the operators hd, where h is a polynomial of degree $< \deg g$ and d an operator in D. Writing an arbitrary operator L

from D as a $\mathbb{C}$-linear combination of monomials ∂^α we deduce that

$$Lg - gL \in g_1 D \text{ for all } L \in D. \tag{5.6.1}$$

More generally, for $k \geq 1$ denote by $g_k D$ the $\mathbb{C}$-vector space generated by the operators hd, where $d \in D$ and h is a polynomial of degree $\leq \deg g - k$. So if $k > \deg g$, then $g_k D = 0$. With these notations we get:

Proposition 5.6.3 *Let $g \in \mathbb{C}[x]$ be a polynomial of degree $d \geq 1$ and $L \in D$. Then for all $k \geq 1$*

$$L^k g \in g L^k + g_1 D L^{k-1} + g_2 D L^{k-2} + \cdots + g_k D.$$

Proof   Follows easily by induction on k: the case $k = 1$ is (5.6.1) and for the general case write $L^{k+1} = L \cdot L^k$ and use that D is a commutative ring.   □

Proof of Theorem 5.6.2   Let f be as in the hypothesis of this theorem and $g \in \mathbb{C}[x]$ of degree $d \geq 1$. Then there exists an N such that $\Lambda^m f^{i+m} = 0$ for all $0 \leq i \leq d$ and all $m \geq N$. By Proposition 5.6.3, using that $g_k D = 0$ if $k > d$, we obtain that

$$\Lambda^m g \in g\Lambda^m + g_1 D\Lambda^{m-1} + g_2 D\Lambda^{m-2} + \cdots + g_d D\Lambda^{m-d}. \tag{5.6.2}$$

If $m \geq N + d$, then $m - i \geq m - d \geq N$, for all $0 \leq i \leq d$. So (5.6.2) and the definition of N give that

$$\Lambda^m g f^m \in g\Lambda^m f^m + g_1 D\Lambda^{m-1} f^m + g_2 D\Lambda^{m-2} f^m + \cdots + g_d D\Lambda^{m-d} f^m = 0$$

which completes the proof.   □

Next we show that it suffices to investigate the Generalized Vanishing Conjecture for a special class of operators. More precisely we show:

Theorem 5.6.4 (de Bondt)   *If the Generalized Vanishing Conjecture holds in all dimensions for all operators of the form*

$$\partial_1^{d_1} + \partial_2^{d_2} + \cdots + \partial_n^{d_n},$$

where the d_i are positive integers, then the Generalized Vanishing Conjecture holds.

Proof

(i) Let $\Lambda \in D$ be a nonzero operator and $f \in \mathbb{C}[x]$ such that $\Lambda^m f^m = 0$ for all $m \geq 1$. By Exercise 1 we may assume that Λ has no constant term. Since each monomial of degree d appearing in Λ can be written as a $\mathbb{Q}$-linear combination of powers of the form l^d, where l is a $\mathbb{Q}$-linear combination of the ∂_i ([117], section 5.2, Exercise 7), we can write Λ as a $\mathbb{C}$-linear combination of such powers, i.e.,

$$\Lambda = c_1 l_1^{d_1} + \cdots + c_N l_N^{d_N} \tag{5.6.3}$$

for some $c_i \in \mathbb{C}^*$, $d_i \geq 1$, where $l_i = a_{i1}\partial_1 + \cdots + a_{in}\partial_n$ for some $a_{ij} \in \mathbb{C}$. Since $\mathbb{C}$ is algebraically closed we can assume that all c_i are equal to 1.

(ii) Now we introduce N new variables $y_1, \ldots, y_N$ and consider the operator

$$\Lambda^* := (\partial_{y_1} + l_1)^{d_1} + \cdots + (\partial_{y_N} + l_N)^{d_N} \tag{5.6.4}$$

on the polynomial ring $\mathbb{C}[x_1, \ldots, x_n, y_1, \ldots, y_N]$. Making the linear coordinate change $x_i' := x_i - (a_{1i}y_1 + \cdots + a_{Ni}y_N)$, $y_j' := y_j$, for all i, j, we obtain that on these new coordinates the operator Λ^* is of the form as described in the statement of the theorem. It follows from the hypothesis that the Generalized Vanishing Conjecture holds for Λ^*.

(iii) Since $f \in \mathbb{C}[x]$ satisfies $\Lambda^m f^m = 0$ for all $m \geq 1$, it follows from (5.6.3) and (5.6.4) and the fact that all c_i are equal to 1 that also $(\Lambda^*)^m f^m = 0$ for all $m \geq 1$. As observed in ii) the Generalized Vanishing Conjecture holds for this operator. So we obtain that for every $g \in \mathbb{C}[x]$ also $(\Lambda^*)^m (gf^m) = 0$ for all large m. But, again by (5.6.3) and (5.6.4), it follows that $\Lambda^m (gf^m) = 0$ for all large m, which concludes the proof. $\square$

Using similar methods we show:

Theorem 5.6.5 (de Bondt)   *Let $\Lambda \in D$ be a product of linear forms (in the ∂_i). Then the Generalized Vanishing Conjecture holds for Λ.*

Proof   Let $\Lambda = l_1 \cdot l_2 \cdots l_N$ for some nonzero linear forms l_i. As above, introduce N new variables $y_1, \ldots, y_N$ and consider the operator

$$\Lambda^* := (\partial_{y_1} + l_1) \cdot (\partial_{y_2} + l_2) \cdots (\partial_{y_N} + l_N).$$

Making the linear coordinate change $x_i' := x_i - (a_{1i}y_1 + \cdots + a_{Ni}y_N)$, $y_j' := y_j$, for all i, j, we obtain that on these new coordinates the operator Λ^* is of the form

$$\partial_{y_1'} \cdot \partial_{y_2'} \cdots \partial_{y_N'}.$$

By Theorem 5.5.6 it follows that the Generalized Vanishing Conjecture holds for this operator. Hence the Generalized Vanishing Conjecture holds for Λ^*. Then, as in iii) of the proof of Theorem 5.6.4, we deduce that the Generalized Vanishing Conjecture holds for Λ.   □

Corollary 5.6.6 *In dimension two the Generalized Vanishing Conjecture holds for any homogeneous operator.*

Proof   Since any homogeneous polynomial in two variables is a product of linear factors, the result follows from Theorem 5.6.5.   □

To conclude this section we describe some more special cases for which the Generalized Vanishing Conjecture holds. Let $0 \neq \Lambda = \Lambda(\partial_1, \ldots, \partial_n) \in D = \mathbb{C}[\partial] = \mathbb{C}[\partial_1, \ldots, \partial_n]$. We define the **polytope of** Λ, denoted $Poly(\Lambda)$, as the polytope of the polynomial $\Lambda(\zeta_1, \ldots, \zeta_n)$.

Proposition 5.6.7 *Let $P \in \mathbb{C}[x]$ and $0 \neq \Lambda \in D$. If*

$$(Poly(P) - Poly(\Lambda)) \cap \mathbb{R}^n_{\geq 0} = \emptyset,$$

then (i) $\Lambda^m(P^m) = 0$ for all $m \geq 1$.
(ii) For every $Q \in \mathbb{C}[x]$: $\Lambda^m(QP^m) = 0$ for all large m.

Proof   Let $\Sigma := Poly(P) - Poly(\Lambda)$. We first prove ii). We may assume that $Q = x^c$ for some $c \in \mathbb{N}^n$. Since $\Sigma \cap \mathbb{R}^n_{\geq 0} = \emptyset$ we have $d := \delta(\Sigma, \mathbb{R}^n_{\geq 0}) > 0$. Hence

$$\delta\big(c + (Poly(P^m) - Poly(\Lambda^m), \mathbb{R}^n_{\geq 0}\big) = \delta(c + m\Sigma, \mathbb{R}^n_{\geq 0}) \geq md - |c| > 0$$

for all large m, say for $m \geq N$. Consequently, if $a \in Supp(P^m) \subseteq Poly(P^m)$ and $b \in Supp(\Lambda^m) \subseteq Poly(\Lambda^m)$, then $c + a - b \notin \mathbb{R}^n_{\geq 0}$. So $\partial^b x^{c+a} = 0$. Since Λ^m is a $\mathbb{C}$-linear combination of the ∂^b, with $b \in Supp(\Lambda^m)$ and P^m a $\mathbb{C}$-linear combination of the x^a, with $a \in Supp(P^m)$, it follows that $\Lambda^m(z^c P^m) = 0$ for all $m \geq N$, which completes the proof of ii). Finally i) follows from the arguments above by choosing $c = 0$ and observing that $\Sigma \cap \mathbb{R}^n_{\geq 0} = \emptyset$ implies that $m\Sigma \cap \mathbb{R}^n_{\geq 0} = \emptyset$ for all $m \geq 1$.   □

Remark 5.6.8 If $\Lambda^m P^m = 0$ for all $m \geq 1$, then the condition in Proposition 5.6.7, $(Poly(P) - Poly(\Lambda)) \cap \mathbb{R}^n_{\geq 0} = \emptyset$, needs not to be satisfied: take $\Lambda = \partial_x - \partial_y$ and $P = x + y$. One easily verifies that $\Lambda^m P^m = 0$ for all $m \geq 1$, however $(0,0) \in (Poly(P) - Poly(\Lambda)) \cap \mathbb{R}^2_{\geq 0}$.

To formulate the next theorem let $e := (e_1, \ldots, e_n) \in \mathbb{Z}^n \setminus \{0\}$. We define a grading ω on $\mathbb{C}[x]$ by defining $\omega(x_i) = e_i$ for all i. So if $a \in \mathbb{N}^n$, then $\omega(x^a) = \langle a, e \rangle = \sum_i a_i e_i$.

Theorem 5.6.9 *Let $a, b \in \mathbb{N}^n$, with $\omega(x^a) \neq \omega(x^b)$, and $\Lambda = c_1 \partial^a + c_2 \partial^b$ with $c_1, c_2 \in \mathbb{C}$. Then the Generalized Vanishing Conjecture holds for Λ and any ω-homogeneous $P \in \mathbb{C}[x]$.*

Proof   Using Theorem 5.5.6 we may assume that both c_i are nonzero. Dividing by c_1 we may assume that $c_1 = 1$. Making a linear coordinate change we may also assume that $c_2 = 1$ (since $a \neq b$) and hence that $\Lambda = \partial^a + \partial^b$. Let $\Sigma := Poly(P) - Poly(\Lambda)$. By Proposition 5.6.7 it suffices to show that $\Sigma \cap \mathbb{R}^n_{\geq 0} = \emptyset$. So assume otherwise. Choose a rational point $w \in \Sigma \cap \mathbb{R}^n_{\geq 0}$. As observed in Sect. 5.5, there exist rational points $u \in Poly(P)$ and $v \in Poly(\Lambda)$ such that $u - v = w \in \mathbb{R}^n_{\geq 0}$, which contradicts Proposition 5.6.10 below.   □

From now on we assume that $\Lambda = \partial^a + \partial^b$ satisfies $\Lambda^m P^m = 0$ for all $m \geq 1$, where $P \in \mathbb{C}[x]$ is an ω- homogeneous polynomial of degree $d \geq 0$.

Proposition 5.6.10 *If $u \in Poly(P)$ and $v \in Poly(\Lambda)$ are rational points, then $u - v \notin \mathbb{R}^n_{\geq 0}$.*

Before we prove this result we make some observations. By the binomial theorem we get

$$0 = \Lambda^m P^m = \sum_{k,l \geq 0, k+l=m} \binom{m}{k} \partial^{ka+lb}(P^m).$$

Claim: all terms in this expression have different degrees (and hence they are all 0). Namely, let $k, l, k', l' \geq 0$ with $k + l = k' + l' = m$. If the terms $\partial^{ka+lb}(P^m)$ and $\partial^{k'a+l'b}(P^m)$ have the same degree, then $k\omega(x^a) + l\omega(x^b) = k'\omega(x^a) + l'\omega(x^b)$, which gives that $(k-k')\omega(x^a) = (l'-l)\omega(x^b)$. Since $k+l = k'+l'$, it follows that $k-k' = l'-l$ and hence, using that $\omega(x^a) \neq \omega(x^b)$, that $k' = k$ and $l' = l$. So $\partial^{ka+lb}(P^m) = 0$ for all $k, l \in \mathbb{N}$ with $k + l = m$.

Since $\Lambda = \partial^a + \partial^b$ the binomial theorem implies that $Supp(\Lambda^m)$ consists of all points of the form $ka + lb$, with $k, l \in \mathbb{N}$ and $k + l = m$. Consequently $\partial^u(P^m) = 0$ for all $u \in Supp(\Lambda)$ and all $m \geq 1$. Write $P^m = \sum_{c \in Supp(P^m)} b_c x^c$ and apply ∂^u, with $u \in Supp(\Lambda)$. This gives that $(c-u)! b_c x^{c-u} = 0$, if $c-u \in \mathbb{R}^n_{\geq 0}$, which is a contradiction. So

$$c - u \notin \mathbb{R}^n_{\geq 0} \text{ for all } c \in Supp(P^m) \text{ and all } u \in Supp(\Lambda^m). \quad (*)$$

Proof of Proposition 5.6.10   Observe that $Poly(\Lambda)$ is the line segment connecting a and b. Since $v, a, b \in \mathbb{Q}^n$ it follows that there exist $r, s \in \mathbb{Q}_{\geq 0}$ with $v = ra+sb$ and $r+s = 1$. Choose $N \in \mathbb{N}, N \geq 1$ such that $Nr, Ns \in \mathbb{N}$. Then, as observed above, $Nv = (Nr)a +$

$(Ns)b \in Supp(\Lambda^N)$, since $Nr+Ns = N$. Furthermore, since $Nu \in Poly(P^N)$, it follows from the Density theorem that there exists $m \geq 1$ such that $mNu \in Supp(P^{Nm})$. Also $mNv \in Supp(\Lambda^{Nm})$, since $Nv \in Supp(\Lambda^N)$. So by (*) we get $mNu - mNv \notin \mathbb{R}^n_{\geq 0}$. Hence $u - v \notin \mathbb{R}^n_{\geq 0}$, which completes the proof. $\qquad\qquad\qquad\qquad\square$

Corollary 5.6.11 *The Generalized Vanishing Conjecture holds for all operators of the form $\partial_x^a + \partial_y^b$ and all homogeneous polynomials in $\mathbb{C}[x, y]$.*

Proof   Follows from Corollary 5.6.6 and Theorem 5.6.9. $\qquad\qquad\qquad\qquad\square$

Exercises for Sect. 5.6

1. Let $\Lambda \in D$ have a nonzero constant term and let $f \in \mathbb{C}[x]$ be such that $\Lambda f = 0$. Show that $f = 0$. Deduce that GVC holds for Λ.

2. For $a \in \mathbb{Z}$ and $m \in \mathbb{N}$, $m \geq 1$ define

$$[a]_m = a(a - 1) \cdots (a - m + 1).$$

   (i) Let $1 \leq r \leq d$. Show that $[2d]_r \geq 2^r[d]_r$.
   (ii) Deduce from (i) that for all $2 \leq r \leq d$: $-4[d]_r[2d - r]_r + [2d]_{2r} \geq 0$.

3. Let $\Lambda = \partial_x - \partial_y^r$, with $r \geq 1$ and k a field of characteristic zero.
   (i) Show that if $0 \neq P \in k[x, y]$ satisfies $\Lambda P = 0$, then $P = e^{x\partial_y^r}(f(y))$ for some $f(y) \in k[y]$.
   (ii) Assume furthermore that $r \geq 2$ and that P is $(r, 1)$-homogeneous such that $\Lambda^2 P^2 = 0$. Deduce that $r > d := deg\, f(y)$ and hence that $P = cy^d$ for some $c \in k^*$ [From i) and the homogeneity of P it follows that $P = ce^{x\partial_y^r}(y^d)$ for some $c \in k^*$. Then use Exercise 2 and the fact that $\Lambda^2 P^2_{|x=0} = 0$].

4. Let $\Lambda_1, \Lambda_2 \in k[\partial]$. Assume that $\Lambda_1(0) \in k^*$. Show that the Generalized Vanishing Conjecture holds for Λ_2 if and only if it holds for $\Lambda_1\Lambda_2$. [Use that Λ_1 is injective on $k[x]$].

5. Let $\omega = (a_1, \ldots, a_n)$, with a_i positive integers and $0 \neq \Lambda(x) \in k[x]$. Let $\Lambda_0(x)$ be the ω-homogeneous part of $\Lambda(x)$ of lowest degree. Put $\Lambda := \Lambda(\partial)$ and $\Lambda_0 := \Lambda_0(\partial)$. Suppose that if $f \in k[x]$ is ω-homogeneous such that $\Lambda_0^m f^m = 0$ for all $m \geq 1$, then $deg\,_\omega\Lambda_0(x) > deg\,_\omega f$. Show that the Generalized Vanishing Conjecture holds for Λ.

6. (Reijnders) Let $\Lambda(x, y) \in k[x, y]$ with $o(\Lambda(x, y)) \leq 1$. Show that the Generalized Vanishing Conjecture holds for $\Lambda := \Lambda(\partial_x, \partial_y)$. [By Exercise 1 we may assume that $o(\Lambda(x, y)) = 1$ and by making a linear coordinate change we may also assume that $\Lambda = \partial_x - \phi(\partial_x, \partial_y)$, with $o(\phi(x, y)) \geq 2$. If $x|\phi(x, y)$ use Exercise 4. If x does not divide $\phi(x, y)$ we can write $\phi(x, y) = xu(x, y) + y^r + v(y)$, with $o(u) \geq 1$ and $o(v) > r$. Let $\Lambda_1 := \sum_{i=0}^r u(\partial_x, \partial_y)^i$ and consider $\Lambda_1\Lambda$. Use Exercises 3, 4 and 5].

Notes

Theorems 5.6.2, 5.6.4, 5.6.5, and Corollary 5.6.6 were obtained by de Bondt in [21]. Proposition 5.6.7 and Theorem 5.6.9 can be found in [128], respectively as Lemma 4.9 and Theorem 6.1. Exercises 2, 3, and 4 are taken from [128]. Finally, Exercises 5 and 6 come from the Master's thesis [101] of Tim Reijnders.

5.7   The Image Conjecture

Let R be a commutative ring and denote by $R[x]$ the polynomial ring in $x_1, \ldots, x_n$ over R.

Image Conjecture (IC(n)) *If R is a $\mathbb{Q}$-algebra and $q \in R[x]$, then $\sum_{i=1}^{n}(\partial_{x_i} - \partial_{x_i}(q))R[x]$ is an MZ-space of $R[x]$.*

This conjecture was formulated by Zhao in [137]. If we take for R the polynomial ring $k[\zeta_1, \ldots, \zeta_n]$, where k is a commutative $\mathbb{Q}$-algebra and $q := \zeta_1 x_1 + \cdots + \zeta_n x_n$ we obtain the Special Image Conjecture described in Sect. 5.1. This follows readily from Proposition 5.2.2. As we have seen in Sect. 5.1, the importance of the Special Image Conjecture comes from the fact that it implies the Generalized Vanishing Conjecture, which in turn implies the Jacobian Conjecture.

What is known about the Special Image Conjecture? As we will show at the end of this section $SIC(1)$ is true. However $SIC(n)$ remains open for all $n \geq 2$. On the other hand, if R is an $\mathbb{F}_p$-*algebra* the analogous problem turns out to have a positive answer, under some mild conditions. In fact we have the following result:

Theorem 5.7.1 *Let R be an $\mathbb{F}_p$-algebra and $a_1, \ldots, a_n$ a regular sequence in R. Then $Im D := \sum_i (\partial_{x_i} - a_i)R[x]$ is an MZ-space of $R[x]$.*

Recall that a sequence $(a_1, \ldots, a_n)$ in R is called a **regular sequence in** R if a_1 is no zero-divisor in R, $R/Ra_1 + \cdots + Ra_{i-1}$ has no a_i-torsion for all $i \geq 2$ and $\mathfrak{a}$, the ideal generated by the a_i, is not equal to R.

Remark 5.7.2 If the a_i do not form a regular sequence the theorem fails: take $n = 1$, $R = \mathbb{F}_p$, and $a_1 = 0$. Then $1 \in \partial_x R[x]$. However $x^{p-1} \notin \partial_x R[x]$. So $\partial_x R[x]$ cannot be an MZ-space by Example 5.2.11.

The proof of Theorem 5.7.1 is based on the following proposition which holds *for every commutative ring R*. We use the notations introduced above.

Proposition 5.7.3 *Let $b \in R[x]$ of degree d and denote by b_d its homogeneous component of degree d. If $b \in Im D$, then all coefficients of b_d belong to the ideal $\mathfrak{a}$.*

Before we prove this result we show how it implies Theorem 5.7.1. First we deduce:

Corollary 5.7.4 *Let $f = \sum f_\alpha x^\alpha$. If $f^p \in Im D$, then $f_\alpha^p \in \mathfrak{a}$ for all α.*

Proof Let $\sum_{i \leq d} f_i$ be the homogeneous decomposition of f. Then $f^p = f_0^p + \cdots + f_d^p$. Since $f^p \in Im D$ it follows from Proposition 5.7.3 that all coefficients of f_d^p belong to $\mathfrak{a}$. So f_d^p is a sum of monomials of the form $c a_i x^{\alpha p}$, with $|\alpha| = d$ and $c \in R$. Since $c a_i x^{\alpha p} = (\partial_i - a_i)(-c x^{\alpha p}) \in Im D$, it follows that $f_d^p \in Im D$. So $f^p - f_d^p \in Im D$. Hence $f_0^p + \cdots + f_{d-1}^p \in Im D$. Then the result follows by induction on d. $\square$

Proof of Theorem 5.7.1 Let $f = \sum f_\alpha x^\alpha$ be such that $f^p \in Im D$. Then $f_\alpha^p \in \mathfrak{a}$ for all α by Corollary 5.7.4. So $f_\alpha^{p^2} \in Ra_1^p + \cdots + Ra_n^p$ for all α. Since $f^{p^2} = \sum f_\alpha^{p^2} x^{p^2 \alpha}$, it follows that for every $g \in R[x]$ all coefficients of $g f^m$ belong to $Ra_1^p + \cdots + Ra_n^p$ if $m \geq p^2$. So all monomials of $g f^m$ are of the form $c a_i^p x^\alpha = (\partial_i - a_i)^p (-c x^\alpha) \in Im D$ (since $\partial_i^p = 0$ on $R[x]$). So $g f^m \in Im D$, which completes the proof. $\square$

So it remains to prove Proposition 5.7.3. We start with the following well-known result about regular sequences:

Lemma 5.7.5 *Let $(a_1, \ldots, a_n)$ be a regular sequence in R. If $g_1, \ldots, g_n \in R$ are such that $\sum a_i g_i = 0$, then for each pair (i, j), with $1 \leq i, j \leq n$ and $i \neq j$, there exists an element $g_{ij} \in R$ such that $g_{ij} = -g_{ji}$ and $g_i = \sum_{j \neq i} g_{ij} a_j$.*

Corollary 5.7.6 *Let $(a_1, \ldots, a_n)$ be a regular sequence in R. If $g_1, \ldots, g_n \in R[x]$ are such that $\sum a_i g_i = 0$, then for each pair (i, j), with $1 \leq i, j \leq n$ and $i \neq j$, there exists an element $g_{ij} \in R[x]$ such that $g_{ij} = -g_{ji}$ and $g_i = \sum_{j \neq i} g_{ij} a_j$.*

Proof Follows immediately from Lemma 5.7.5 by looking at each monomial x^α in the equation $\sum a_i g_i = 0$. $\square$

Proof of Proposition 5.7.3 Let $b \in Im D$ and denote by b_d its highest degree homogeneous component. So

$$b = \sum_{i=1}^{n} (\partial_i - a_i) h^{(i)} \tag{5.7.1}$$

for some $h^{(i)} \in R[x]$ and at least one $h^{(i)}$ has degree $\geq d$. If all $h^{(i)}$ have degree $\leq d$ it follows from (5.7.1) that $b_d = \sum_{i=1}^{n} (-a_i) h_d^{(i)}$, which gives the desired result. So we may assume that for some $m \geq 1$ all $h^{(i)}$ have the following homogeneous decomposition:

$$h^{(i)} = h_{d+m}^{(i)} + h_{d+m-1}^{(i)} + \cdots$$

Now we claim: for each $1 \leq k \leq m$ and each pair (i, j) with $1 \leq i, j \leq n$ and $i \neq j$, there exists a polynomial $(g_{d+k})_{ij} \in R[x]$, homogeneous of degree $d + k$, such that $(g_{d+k})_{ij} = -(g_{d+k})_{ji}$ and

$$h_{d+k}^{(i)} = \sum_{j \neq i} (g_{d+k})_{ij} a_j - \sum_{j \neq i} \partial_j (g_{d+k+1})_{ij}$$

and furthermore $(g_{d+m+1})_{ij} = 0$ for all $i \neq j$.

To prove this claim we use inverse induction on k, starting with $k = m$. Therefore consider the homogeneous part of degree $d + m$ in (5.7.1). This gives $\sum_i a_i h_{d+m}^{(i)} = 0$. So by Corollary 5.7.6 there exist $(g_{d+m})_{ij} \in R[x]$, homogeneous of degree $d + m$, such that $(g_{d+m})_{ij} = -(g_{d+m})_{ji}$ and

$$h_{d+m}^{(i)} = \sum_{j \neq i} (g_{d+m})_{ij} a_j.$$

This proves the case $k = m$, since $(g_{d+m+1})_{ij} = 0$ for all $i \neq j$.

Now let $1 \leq k \leq m - 1$. By the induction hypothesis we know the case $k + 1$. So there exist $(g_{d+k+1})_{ij} \in R[x]$ such that $(g_{d+k+1})_{ij} = -(g_{d+k+1})_{ji}$ and

$$h_{d+k+1}^{(i)} = \sum_{j \neq i} (g_{d+k+1})_{ij} a_j - \sum_{j \neq i} \partial_j (g_{d+k+2})_{ij}. \tag{5.7.2}$$

Now look at the homogeneous part of degree $d + k$ in (5.7.1). This gives

$$\sum_{i=1}^{n} (-a_i) h_{d+k}^{(i)} + \sum_{i=1}^{n} \partial_i h_{d+k+1}^{(i)} = 0.$$

Substituting (5.7.2) into this equation gives

$$\sum_{i=1}^{n} (-a_i) h_{d+k}^{(i)} + \sum_{i=1}^{n} \partial_i \Big[\sum_{j \neq i} (g_{d+k+1})_{ij} a_j \Big] - \sum_{i=1}^{n} \partial_i \Big[\sum_{j \neq i} \partial_j (g_{d+k+2})_{ij} \Big] = 0. \tag{5.7.3}$$

Now observe that the last sum is equal to zero, since

$$\partial_i \partial_j (g_{d+k+2})_{ij} + \partial_j \partial_i (g_{d+k+2})_{ji} = 0$$

because $(g_{d+k+2})_{ij} = -(g_{d+k+2})_{ji}$. The second sum equals

$$\sum_{s=1}^{n} \partial_s \Big[\sum_{j \neq s} (g_{d+k+1})_{sj} a_j \Big].$$

So the a_i-part of this expression equals

$$\sum_{s=1,s\neq i}^{n} \partial_s (g_{d+k+1})_{si} a_i.$$

So (5.7.3) becomes

$$-\sum_{i=1}^{n} a_i \left[h_{d+k}^{(i)} - \sum_{s=1,s\neq i}^{n} \partial_s (g_{d+k+1})_{si} \right] = 0.$$

Hence, replacing s by j, we get

$$\sum_{i=1}^{n} a_i \left[h_{d+k}^{(i)} - \sum_{j\neq i} \partial_j (g_{d+k+1})_{ji} \right] = 0.$$

So by Corollary 5.7.6 there exist $(g_{d+k})_{ij} \in R[x]$, homogeneous of degree $d+k$ such that $(g_{d+k})_{ij} = -(g_{d+k})_{ji}$ and

$$h_{d+k}^{(i)} - \sum_{j\neq i} \partial_j (g_{d+k+1})_{ji} = \sum_{j\neq i} (g_{d+k})_{ij} a_j.$$

So

$$h_{d+k}^{(i)} = \sum_{j\neq i} (g_{d+k})_{ij} a_j + \sum_{j\neq i} \partial_j (g_{d+k+1})_{ji}.$$

Since $(g_{d+k+1})_{ji} = -(g_{d+k+1})_{ij}$ the claim follows.

Now we can finish the proof: in particular for $k = 1$ we get

$$h_{d+1}^{(i)} = \sum_{j\neq i} (g_{d+1})_{ij} a_j + \sum_{j\neq i} \partial_j (g_{d+2})_{ji}. \tag{5.7.4}$$

Furthermore, from (5.7.1) we get

$$b_d = \sum_{i=1}^{n} \partial_i h_{d+1}^{(i)} - \sum_{i=1}^{n} a_i h_d^{(i)}.$$

Finally, substituting (5.7.4) into the last equation and using, as before, that $\sum_{i=1}^{n} \partial_i \left[\sum_{j\neq i}^{n} \partial_j (g_{d+2})_{ji} \right] = 0$, we obtain the desired result. $\qquad \square$

As an immediate consequence of Theorem 5.7.1, taking for R the polynomial ring $k[\zeta_1, \ldots, \zeta_n]$, we get:

Corollary 5.7.7 *Let k be an $\mathbb{F}_p$-algebra and $D := \{\partial_{x_1} - \zeta_1, \ldots, \partial_{x_n} - \zeta_n\}$ on $k[x, \zeta]$. Then $Im D$ is an MZ-space of $k[x, \zeta]$.*

To conclude this section we first show that $SIC(1)$ holds and finally in Theorem 5.7.10 that $IC(1)$ holds for the operator $\partial_x - a$, where $a \in R$, in case R is a UFD.

Theorem 5.7.8 *Let k be a commutative ring contained in a $\mathbb{Q}$-algebra and $D := \partial_x - \zeta$. Then $Im D := (\partial_x - \zeta)k[x, \zeta]$ is an MZ-space of $k[x, \zeta]$.*

Proof (Started) Let $\mathcal{E}$ be the k-linear map from $A := k[x, \zeta]$ to $k[x]$ defined by $\mathcal{E}(\zeta^a x^b) = \partial^a(x^b)$. By Proposition 5.1.11 $Im D = \ker \mathcal{E}$. So we need to show that $\ker \mathcal{E}$ is an MZ-space of A. Therefore first observe that $\mathcal{E}(\zeta^a x^b) = 0$ if $a > b$. Furthermore, for any nonzero polynomial g in A we define its Degree, denoted $Deg(g)$, as the maximum of the Degrees of all nonzero monomials appearing in g, where $Deg(c \zeta^a x^b) = b - a$ if $c \in k \backslash \{0\}$. It follows that

$$\text{if } Deg(g) \leq -1, \text{ then } \mathcal{E}(g) = 0. \tag{5.7.5}$$

In particular, if for some element f of A its Degree is ≤ -1, then for all $m \geq 1$ the Degree of all powers f^m are ≤ -1. So by (5.7.5) we get that $\mathcal{E}(f^m) = 0$ for all $m \geq 1$. Below we will show that the converse is true as well, if k is a reduced ring, i.e., if its nilradical $\mathfrak{n}$ equals zero. More precisely: □

Proposition 5.7.9 *Let k be a reduced ring and $f \in A$. Then $\mathcal{E}(f^m) = 0$ for all large m if and only if $Deg(f) \leq -1$.*

Proof of Theorem 5.7.8 (Completed) We first show how Proposition 5.7.9 implies Theorem 5.7.8. First we may assume that k is a $\mathbb{Q}$-algebra. Consequently also $\bar{k} := k/\mathfrak{n}$ is a $\mathbb{Q}$-algebra. Furthermore it is a reduced ring. Now assume that $\mathcal{E}(f^m) = 0$ for all large m. Then $\mathcal{E}(\bar{f}^m) = 0$ for all large m, where $\bar{f}$ is the element in $\bar{k}[x, \zeta]$ obtained by reducing the coefficients of f mod $\mathfrak{n}$. So by the proposition $Deg(\bar{f}) \leq -1$. Hence $f = f_* + f_-$, where f_* is nilpotent and $Deg(f_-) \leq -1$. So there exists $r \geq 1$ such that $f_*^r = 0$. Now let g in A be nonzero and let d be its Degree. Then for large m, using that $f_*^r = 0$, we get

$$g f^m = \sum_{k=0}^{r-1} \binom{m}{k} g f_*^k f_-^{m-k}.$$

Let $Deg(f_*) \leq e$. Then $Deg(gf_*^k) \leq d + (r-1)e$ for all $0 \leq k \leq r-1$. Furthermore if m is large, then for all $0 \leq k \leq r-1$ we have that $m-k > d+(r-1)e$. Consequently for such k we have

$$Deg(gf_*^k f_-^{m-k}) \leq -1 \text{ for all large } m. \tag{5.7.6}$$

Then by (5.7.5) and (5.7.6) we get $\mathscr{E}(gf^m) = 0$ for all large m, which completes the proof of Theorem 5.7.8.   □

To prove Proposition 5.7.9, let f be such that $\mathscr{E}(f^m) = 0$ for all large m. *Assume that* $r := Deg(f) \geq 0$. We will deduce a contradiction. Namely, let $g := \zeta^r f$. Then $Deg(g) = 0$. Furthermore

$$\mathscr{E}(g^m) = \mathscr{E}(\zeta^{mr} f^m) = \partial^{mr} \mathscr{E}(f^m) = 0 \tag{5.7.7}$$

for all large m. Writing g in its homogeneous decomposition, using that the Degree of g is zero, we get that $g = g_0 + g_{-1} + \cdots$ and hence that $g^m = g_0^m + g_-$, where g_0 is nonzero and g_- has Degree ≤ -1. Applying $\mathscr{E}$ to the last equality it follows from (5.7.5) and (5.7.7) that $\mathscr{E}(g_0^m) = \mathscr{E}(g^m) = 0$ for all large m. Summarizing, if $r \geq 0$ there exists an nonzero element g_0, which is homogeneous of Degree zero, such that $\mathscr{E}(g_0^m) = 0$ for all large m.

Now write g instead of g_0. Since g is homogeneous of Degree zero it is a sum of monomials of the form $c_a \zeta^a x^a$. In other words $g = P(u)$, a nonzero polynomial in $u := \zeta x$ over k. Now observe that $\mathscr{E}(u^n) = \mathscr{E}(\zeta^n x^n) = n!$ and that by Corollary 5.4.14 the Factorial Conjecture holds in dimension one. Consequently, since $\mathscr{E}(P(u)^m) = \mathscr{E}(g^m) = 0$ for all large m, we get that $P(u) = 0$ (since k is reduced!), i.e., $g = 0$, a contradiction since $g = g_0$ is nonzero.

In the remainder of this section R is a $\mathbb{Q}$-algebra and $R[t]$ denotes the univariate polynomial ring over R.

Theorem 5.7.10 *Let R be a UFD. For any $a \in R$ set $D := \partial_t - a$. Then $Im\,D$ is an MZ-space of $R[t]$.*

The proof of this theorem is based on the next lemma. First, as in Exercise 1 of Sect. 5.2, let $\mathfrak{L} : R[t] \to R$ denote the R-linear map defined by $\mathfrak{L}(t^n) = n!$ for all $n \geq 0$. For an element a in R we denote by $r(a)$ the radical ideal of Ra. If a is *not a zero-divisor* in R, then it is easy to see that

$$\text{for any } b \in R,\ b \in Im\,D \text{ if and only if } b \in Ra. \tag{**}$$

Lemma 5.7.11 *Notations as above. Assume that a is not a zero-divisor in R. Then the following statements hold:*

(i) *For every $n \geq 0$ $a^n t^n \equiv n! (\mathrm{mod}\ Im D)$. Furthermore, $R a^{n+1} t^n \subseteq Im D$ for all $n \geq 0$.*

(ii) *Let $f = p(at)$, for some $p(t) \in R[t]$. Then $f \in Im D$ if and only if $\mathfrak{L}(p) \in Ra$.*

(iii) *Let f be as in (ii). Then $f \in r(Im D)$ if and only if all coefficients of $p(t)$ belong to $r(a)$.*

Proof

(i) The first statement follows by induction on n from the equality $D(a^{n-1} t^n) = n a^{n-1} t^{n-1} - a^n t^n$. The second statement follows by multiplying the first one by a and using the equivalence in (**).

(ii) By (i) and the definition of the R-linear map $\mathfrak{L}$ we have that $f = p(at) \in Im D$ if and only if $\mathfrak{L}(p) \in Im D$. The statement then follows from the equivalence in (**).

(iii) Write $p(t) = \sum c_i t^i$, with $c_i \in R$. It follows from (ii) that $f \in r(Im D)$ if and only if $\mathfrak{L}(p^m) \in Ra$ for all large m. Clearly, if all c_i belong to $r(a)$, then for all large m we have that $\mathfrak{L}(p^m)$ belongs to Ra, whence $f \in r(Im D)$.

Conversely, assume that $\mathfrak{L}(p^m) \in Ra$ for all large m. Let $\mathfrak{p}$ be a prime ideal of R containing Ra and denote by K the quotient field of $R/\mathfrak{p}$. Reducing p modulo $\mathfrak{p}$ we get the polynomial $\overline{p}$, which we view inside $K[t]$. Then $\mathfrak{L}(p^m) \in Ra$ for all large m implies that $\mathfrak{L}_1(\overline{p}^m) = 0$ for all large m, where $\mathfrak{L}_1 : K[t] \to K$ is the K-linear map defined by $\mathfrak{L}_1(t^n) = n!$ for all $n \geq 0$. It then follows from Corollary 5.4.14 that $\overline{p} = 0$. So all coefficients of p belong to $\mathfrak{p}$. Since this holds for all primes $\mathfrak{p}$ containing Ra and $r(a)$ is the intersection of all such primes, we get that all coefficients of p belong to $r(a)$. $\qquad\square$

Corollary 5.7.12 *Let $f = p(at)$ for some $p \in R[t]$. If $f \in r(Im D)$, then for every $g \in R[t]$ we have that $g f^m \in Im D$ for all large m.*

Proof We may assume that $g = t^d$ for some $d \geq 0$. By Lemma 5.7.11 all coefficients of $p(t)$ belong to $r(a)$. So there exists $N \geq 1$ such that all coefficients of $p(t)^N$ belong to Ra. Hence all coefficients of $p(t)^{N(d+1)}$ belong to Ra^{d+1} and the same holds for all coefficients of $p(t)^m$, whenever $m \geq N(d+1)$. So for such m and each $i \geq 0$ the coefficient of t^i in $f^m = p(at)^m$ belongs to Ra^{i+d+1}. Consequently, for each $n \geq 0$ the coefficient of t^n in $t^d f^m$ belongs to Ra^{n+1}. It then follows from Lemma 5.7.11 (i) that $t^d f^m \in Im D$ for all $m \geq N(d+1)$, which completes the proof. $\qquad\square$

Proof of Theorem 5.7.10 First, if $a = 0$ we are done. So assume $a \neq 0$. Furthermore, if a is a unit in R, then, since ∂_t is locally nilpotent on $R[t]$, the R-linear operator D is invertible with inverse given by $D^{-1} = -\sum_{i \geq 0} a^{-i-1} \partial_t^i$. Hence $Im D = R[t]$, which is an MZ-space of $R[t]$.

So we may assume that Ra is a proper ideal in R. Hence so is the ideal $I := \bigcap_{i=0}^{\infty} Ra^i$. Then for each $c \in R \setminus I$ there exists a unique integer $n \geq 0$ such that $c \in Ra^n$ but $c \notin Ra^{n+1}$. We define $v_a(c) := n$ in this case and set $v_a(c) := \infty$ for all $c \in I$. Now let $f \in r(Im D)$ and write $f = \sum_{i=0}^{d} c_i t^i$, with $c_i \in R$. If all coefficients c_i belong to I, then f can be written in the form $p(at)$ for some $p(t) \in R[t]$. Then by Corollary 5.7.12 for every $g \in R[t]$ we have that $g f^m \in Im D$ for all large m. So we may assume that not all c_i belong to I.

Let $s(f)$ be the minimum of all $v_a(c_i) - i$, with $0 \leq i \leq d$. If $s(f) \geq 0$, then again by Corollary 5.7.12 we are done. So assume that $s(f) \leq -1$. We will derive a contradiction, which completes the proof. Therefore put $f_1 := a^{-s(f)} f$. Then $f_1 \in r(Im D)$ and $s(f_1) = 0$. Write $f_1 = \sum b_i t^i$. Then each b_i is of the form $b_i = a^i d_i$ for some $d_i \in R$ and furthermore $d_i \notin Ra$ for some i. Let $p(t) := \sum d_i t^i$. Then $f_1(t) = p(at)$. By Lemma 5.7.11 (iii) we get that $d_i \in r(a)$ for all i. By Lemma 5.7.13 below there exist $u \in R$ and $e_i \in R$ such that $u d_i = e_i a$ for all i and $e_i \notin r(a)$ for some i. Since $-s(f) \geq 1$ we get that $a^{-s(f)-1} \in R$ and hence that $a^{-s(f)-1} f \in r(Im D)$. Hence so does the polynomial $f_2 := u a^{-s(f)-1} f = u a^{-1} f_1$. Now observe that

$$f_2 = u a^{-1} f_1 = u a^{-1} \sum b_i t^i = a^{-1} \sum u d_i a^i t^i = \sum e_i a^i t^i.$$

Hence $f_2(t) = q(at)$, where $q(t) = \sum e_i t^i \in R[t]$. Then by Lemma 5.7.11 (iii), since $f_2 \in r(Im D)$, we get that all e_i belong to $r(a)$. But this a contradiction, since, as pointed out above, $e_i \notin r(a)$ for some i.   □

Lemma 5.7.13 *Let R be a UFD. If $d_1, \ldots, d_n \in r(a)$ are such that $d_i \notin Ra$ for some i, then there exist elements $u \in R$ and $e_i \in R$ such that $u d_i = e_i a$ for all i and $e_i \notin r(a)$ for some i.*

Proof Let g be the greatest common divisor of a and all d_i. Put $u := a/g$ and $e_i := d_i/g$ for all i. Since for some i $d_i \notin Ra$ there exists a prime factor p of a such that $v_p(a)$, the multiplicity of p in a, is larger than $v_p(d_i)$, the multiplicity of p in d_i. Choose i such that $v_p(d_i)$ is minimal. For such an i we have that $v_p(e_i) = 0$. Consequently $e_i \notin r(a)$. Finally $u d_i = a \cdot d_i/g = a e_i$, which completes the proof.   □

Corollary 5.7.14 *For each $1 \leq i \leq n$ the image $(\partial_{x_i} - \zeta_i) \mathbb{C}[x, \zeta]$ is an MZ-space of $\mathbb{C}[x, \zeta]$.*

Proof Apply Theorem 5.7.10 to $R = \mathbb{C}[\zeta][x_1, \ldots, x_{i-1}, x_{i+1}, \ldots, x_n]$, $t = x_i$ and $a = \zeta_i$.   □

Remark 5.7.15 Since $SIC(n)$ asserts that $\sum_{i=1}^{n} (\partial_{x_i} - \zeta_i) \mathbb{C}[x, \zeta]$ is an MZ-space of $\mathbb{C}[x, \zeta]$ and by Corollary 5.7.14 each summand is an MZ-space of $\mathbb{C}[x, \zeta]$, one could hope that the sum of two MZ-spaces (and hence the sum of a finite number of these spaces) is again an MZ-space. However this is not true as can be seen from Exercise 1 below.

Remark 5.7.16 If R is a domain which is not a UFD, then the statement of Lemma 5.7.13 does not hold. This can be seen from Exercise 2 below.

Exercises for Sect. 5.7

1. Show that the sum of two MZ-spaces of $\mathbb{C}[x]$ need not be an MZ-space [Take $M_1 = \mathbb{C}x_1, M_2 = \mathbb{C}(1 - x_1)$ and use Example 5.2.11].
2. Show that there exists a $\mathbb{Q}$-algebra which is a domain, but does not satisfy the statement of 5.7.13. [Take $R = \mathbb{Q}[t^2, t^3], a = t^2, d_1 = t^3$ and $d_i = a$ for all $i > 1$].
3. Let R be a $\mathbb{Q}$-algebra and $a \in R$ a nonzero divisor. Put $D := \partial_t - a$. Show that $f = \sum_{i=0}^d c_i t^i \in R[t]$ belongs to $Im D$ if and only if

$$d! c_d + (d - 1)! c_{d-1} a + (d - 2)! c_{d-2} a^2 + \cdots + c_0 a^d \equiv 0 \, (mod \, a^{d+1}).$$

[Use induction on d and the fact that $c_d = -ab_d$ for some $b_d \in R$].

Notes
Theorem 5.7.1 is Theorem 2.2 of [127]. Also the proof given here is taken from that paper. Theorem 5.7.8 is taken from the paper [118]. Finally, Theorem 5.7.10 is Theorem 7.1 in [125]. Also its proof comes from [125].

5.8   MZ-Spaces of Matrices of Codimension One

In this section we give a complete classification of all MZ-spaces of codimension 1 of rings which are finite products of matrix rings over a field k. We use the following notation: k is a field and $A = A_1 \times \cdots \times A_s$, where $A_i = M_{n_i}(k)$ is the ring of $n_i \times n_i$ matrices over k. First we need some simple observations.

Let $c, a \in M_n(k)$. Then $Tr(c^t a) = \sum_{i,j} a_{ij} c_{ij}$. So if V is a k-linear subspace of $M_n(k)$ of codimension 1, then there exists $0 \neq c \in M_n(k)$ such that $V = \{a \in M_n(k) \mid Tr(ca) = 0\}$. More generally, every codimension 1 subspace of A can be written in the form

$$V = \{(a_1, \ldots, a_s) \in M_n(k)^s \mid \sum Tr(c_i a_i) = 0\} \quad (*)$$

for some $c_i \in A_i$, not all zero. We may assume that for some $1 \leq t \leq s$ the first t c_i's are nonzero and all other c_i are zero. From Exercise 1 it follows that $V = W \times A_{t+1} \times \cdots \times A_s$, where $W = \pi(V)$ and $\pi : A \to A_1 \times \cdots \times A_t$ is the projection map. Also it follows from Exercise 1 that V is an MZ-space of A if and only if W is an MZ-space of $A_1 \times \cdots \times A_t$. Summarizing: in order to classify all MZ-spaces of A we may assume that all c_i are nonzero. Therefore the following result gives a complete classification of all MZ-spaces of A of codimension 1.

Theorem 5.8.1 (Konijnenberg) *Let V be as in (*) with all $c_i \neq 0$. Then V is an MZ-space of A (of codimension 1) if and only if the following conditions hold:*

(i) There exist $\lambda_1, \ldots, \lambda_s \in k$ such that $c_i = \lambda_i I_{n_i}$ for all i.
(ii) $d_1 \lambda_1 + \cdots + d_s \lambda_s \neq 0$ for all $0 \neq d = (d_1, \ldots, d_s)$, where $d_i \in \{0, 1, \ldots, n_i\}$.

Proof ($\Leftarrow$) We show that if $e = (e_1, \ldots, e_s) \in V$ is an idempotent, then $e = 0$. Then it follows from Theorem 5.3.1 that V is an MZ-space of A. So let $e \in V$. Observe that $Tr\, e_i = d_i \in \{0, 1, \ldots, n_i\}$ (since an idempotent matrix can be conjugated to a diagonal matrix with only 0 and 1 on the diagonal). Since $e \in V$ we get $\sum \lambda_i d_i = 0$. Hence by (ii) all d_i are equal to 0 and hence all e_i are 0. So $e = 0$.

($\Rightarrow$) Assume $c_1 \notin k I_{n_1}$. Then by Lemma 5.8.2 below there exists $e_1 \in A_1$ with $e_1^2 = e_1$, $c_1 e_1 \neq 0$ and $Tr\,(c_1 e_1) = 0$. Let $e = (e_1, 0, \ldots, 0)$. Then $e^2 = e$ and $e \in V$. Since V is an MZ-space it follows that $cea \in V$ for all $a \in A$. So $Tr\,(c_1 e_1 a_1) = 0$ for all $a_1 \in A_1$, which implies that $c_1 e_1 = 0$, a contradiction. So $c_1 \in k I_{n_1}$ and similarly $c_i \in k I_{n_i}$ for all i, say $c_i = \lambda_i I_{n_i}$ with $\lambda_i \in k$.

Finally, let $(d_1, \ldots, d_s) \neq 0$ with $d_i \in \{0, \ldots, n_i\}$ and suppose that $\sum d_i \lambda_i = 0$. Let $e_i \in A_i$ be the diagonal matrix whose first d_i diagonal elements are equal to 1 and the others are zero. Then $e_i^2 = e_i$ and $Tr\, e_i = d_i$. Put $e = (e_1, \ldots, e_s)$. Since there exists an i with $d_i \neq 0$, the matrix e_i is nonzero, hence e is a nonzero idempotent. Furthermore $\sum \lambda_i Tr\, e_i = \sum \lambda_i d_i = 0$. So $e \in V$. Since V is an MZ-space of A this implies that $(0, \ldots, 0, a_i, 0, \ldots, 0) \cdot e \cdot (0, \ldots, 0, b_i, 0, \ldots, 0) \in V$ for all $a_i, b_i \in A_i$. Since A_i is a simple ring we get that $(0, \ldots, 0, A_i, 0, \ldots, 0) \subseteq V$. This gives that $Tr(c_i a_i) = 0$ for all $a_i \in A_i$. Hence $c_i = 0$, a contradiction.   □

Lemma 5.8.2 *If $c \notin k I_n$ there exists $e \in M_n(k)$ with $e^2 = e$, $ce \neq 0$ and $Tr\,(ce) = 0$.*

Proof Let $c \notin k I_n$. Then there exists $0 \neq v \in k^n$ which is not an eigenvector of c. So $cv \neq 0$ and $v \notin kcv$. Hence there exists $w \in k^n$ with $w^t cv = 0$ and $w^t v = 1$. Let $e = vw^t$. Then $Tr\,(ce) = Tr\,(cv)w^t = Tr\, w^t cv = 0$, since $w^t cv = 0$. Furthermore $e^2 = vw^t vw^t = v \cdot 1 \cdot w^t = e$. Finally, $ce = cvw^t \neq 0$, for if $ce = 0$ then $cvw^t v = 0$, i.e., $cv = 0$, a contradiction.   □

Corollary 5.8.3 *Let $n := max_i\, n_i$*

(i) If k has characteristic $p \leq n$, there do not exist MZ-spaces of A of codimension 1.
(ii) If the characteristic of k is either zero or $p > n$, there do exist MZ-spaces of A of codimension 1.

Proof

(i) If $p \leq n$, there exists i such that $p \leq n_i$. Let us say for $i = 1$. Then choose $d = (p, 0, \ldots, 0)$. Then for all $\lambda_i \in k$ we have $\sum \lambda_i d_i = 0$ in k. So by Theorem 5.8.1 there cannot exist an MZ-space of A of codimension 1.
(ii) Take for example $V = W \times A_2 \times \cdots \times A_s$, where W is the set of trace zero matrices of A_1.   □

Remark 5.8.4 Let k be algebraically closed. Then the Artin–Wedderburn theorem implies that every finite dimensional semisimple k-algebra is isomorphic to a finite product of matrix rings over k. So by Theorem 5.8.1 we obtain a complete classification of all codimension 1 MZ-spaces of such rings. In particular, if G is a finite group such that the characteristic of k does not divide the order of the group G, then by Maschke's theorem $k[G]$ is semisimple, hence we also understand all codimension 1 MZ-spaces of such group rings.

Exercises for Sect. 5.8

1. Let V be as in (*), $1 \leq t \leq s$ and $c_{t+1} = \cdots = c_s = 0$. Let $\pi : A_1 \times \cdots \times A_s \rightarrow A_1 \times \cdots \times A_t$ be the projection map and define $W = \pi(V)$.
   (i) Show that $V = W \times A_{t+1} \times \cdots \times A_s$.
   (ii) Show that V is an MZ-space of $A_1 \times \cdots \times A_s$ if and only if W is an MZ-space of $A_1 \times \cdots \times A_t$ [Use Theorem 5.2.19].
2. Let k be a field, $n \geq 3$ and $V \subseteq M_n(k)$ a k-linear subspace of codimension two. Let

$$C(V) := \{C \in M_n(k) \mid Tr(CA) = 0 \text{ for all } A \in V\}.$$

   (i) Show that there exist $C_1, C_2 \in C(V)$, linearly independent over k, such that $C(V) = kC_1 + kC_2$.
   (ii) Let $T \in GL_n(k)$. Show that $C(T^{-1}VT) = T^{-1}C(V)T$.
3. Let $n \geq 3$ and $C \in M_n(k)$, not a diagonal matrix. Show that there exists a $T \in GL_n(k)$ such that the second row of $T^{-1}CT$ has the form $(p, *, 0, *, \ldots, *)$ and that its third row has the form $(q, 0, *, 0, \ldots, 0)$ with $pq \neq 0$.
4. (i) Show that every rank 1 matrix in $M_n(k)$ is of the form vw^t for some $v, w \in k^n$.
   (ii) Let $E := vw^t$ have rank 1. Show that $E^2 = E$ if and only if $w^t v (= Tr E) = 1$.
5. (de Bondt) Let $V \subseteq M_n(k)$ be a k-linear subspace. Show that V contains no idempotent of rank 1 if and only if for every $v \in k^n$ there exists a $C \in C(V)$ such that $Cv = v$. [For ($\leftarrow$) use ii) of Exercise 4. For ($\rightarrow$) assume that there exists a $v \in V$ such that $Cv \neq v$ for all $C \in C(V)$. Then $v \notin \sum_{C \in C(V)} kCv$. Choose $w \in k^n$ with $w^t v = 1$ and $w^t Cv = 0$ for all $C \in C(V)$. Then look at $E := vw^t$].
6. Let $V \subseteq M_n(k)$ be a k-linear subspace. Show that V contains no idempotent of rank 1 if and only if for every $v \in k^n$ there exists a $C \in C(V)$ such that $C^t v = v$. [Use Exercise 5 and the fact that $Tr(C^t A^t) = Tr(CA)$].

7. (Konijnenberg) Let $n \geq 3$ and $V \subseteq M_n(k)$ a k-linear subspace of dimension $n^2 - 2$. Show that if V has no idempotent of rank 1, then $I_n \in C(V)$, i.e., V is a subset of the matrices having trace zero. (The proof outlined below is due to M. de Bondt).

   (i) If every element of $C(V)$ is a diagonal matrix, show that $I_n \in C(V)$ [Use Exercise 5 with $v = (1, 1, \ldots, 1)^t$].

   (ii) From now on assume that there exists a non-diagonal matrix $C_1 \in C(V)$. Show that we may assume that C_1 has its second row and its third row as in Exercise 3.

   (iii) Choose $C_2 \in C(V)$ such that $C_2 e_1 = e_1$ and deduce that $C(V) = kC_1 + kC_2$.

   (iv) In the remainder of this exercise we will prove that $C_2 = I_n$. Show that if $C_2(e_1 + e_i) = e_1 + e_i$ for all $i \neq 1$, then $C_2 = I_n$.

   (v) Let $i \neq 3$. Show that $(C_2)_{3i} = 0$ [By Exercise 6 choose $C \in C(V)$ with $C^t e_3 = e_3$. Write $C = aC_1 + bC_2$, $a, b \in k$ and look at the third component of $C^t e_3 = e_3$].

   (vi) Let $i \neq 3$. Show that $C_2(e_1 + e_i) = d(e_1 + e_i)$ for some $d \in k^*$ [By Exercise 5 choose $C_0 \in C(V)$ with $C_0(e_1 + e_i) = e_1 + e_i$ and look at the third component of this equation by writing $C_0 = rC_1 + sC_2$, $r, s \in k$].

   (vii) Let $i \neq 1$, $i \neq 3$. Show that $d = 1$ [By Exercise 5 choose $C_* \in C(V)$ with $C_*((1 - d)e_1 + e_i) = (1 - d)e_1 + e_i$ and look at the third component of this equation by writing $C_* = uC_1 + vC_2$, $u, v \in k$].

   (viii) Show that $C_2(e_1 + e_3) = e_1 + e_3$ by copying the arguments above, replacing "third" by "second" and i by 3.

   (ix) Deduce from (vi), (vii), (viii), and (iv) that $I_n \in C(V)$.

Notes

The results in this section are all taken from the Master's thesis of Aart Konijnenberg [65]. In his thesis he also classifies all two-sided MZ-spaces of rings which are products of matrix rings $M_{n_i}(k)$, where all $n_i \geq 3$ and k is a field of characteristic zero. This result is based on a classification of all codimension ≤ 2 MZ-spaces of $M_n(k)$ in case $n \geq 3$ (see for more details Exercises 2–7). This result of Konijnenberg is in turn extensively generalized by de Bondt: in his paper [23] he shows (amongst other results) that if k has characteristic zero and V is a (left) MZ-space of $M_n(k)$ of codimension less than n, then V is a subspace of the set of matrices having trace zero. Conversely, by the examples 5.2.3 and 5.2.4 each subspace of the set of trace zero matrices is a (left) MZ-space of $M_n(k)$.

Some Corrections to [117]

Below is a list of errors and misprints discovered in [117] during a period of almost 20 years. Certainly, this list will not be complete, but hopefully it will be helpful. I like to thank Gene Freudenburg and Takanori Nagamine for sending me various misprints.

page 5, line 4: $(\bmod\ \mathfrak{m}^2)$

page 19, line 10: $D = 0$ is also allowed

page 21, line -13: characteristic zero

page 32, line 8: **Proof:** Let $D = a\partial_1 + b\partial_2$ and $g = \cdots$

page 34, line -4: $b_1, b2 \in B$ are nonzero

page 35, line -10: if $n \geq 2$

page 41, line 5: no comma after |

page 41, line 11: $k(x_1, \ldots, x_{t-1})(a_i)$

page 58, line -7: $^T G$

page 61, line -2: $R[[X, Y]]$

page 67, line -2: $K[X, Y]$

page 67, line 15: So $B_j \cap K[Y]$

page 78, line 10: has no solution in V

page 86, line 18: $J(R, 2)$

page 96, line -17: $E_1(g) \circ E_1(h)$

page 103, line 21: Example 5.5

page 104, line -10: do not satisfy

page 107, line 7: $K[f]_n = \sum_{i=0}^{n}$

page 107, line 8: $A[f]_n = \sum_{i=0}^{n}$

page 108, line 12: $\deg S := \{\deg s \mid s \in S\}$

page 108, line 21: for all $1 \leq i \leq n$

page 109, line -8: $(i - j)d_{c(s)} \in (d_0, d_m, d_n) \subseteq D_{s-1}$

page 110, line -14: $B_s \subseteq K[f]_{e_s c(s)-1}$

page 111, line 6: $B_s = \{b_1, \ldots, b_r\}$

page 111, line 10: $X^{(e-1)n+n-1}$

page 113, line -4: $H = (1, b_1, \ldots, b_{N-1})$

© The Author(s), under exclusive license to Springer Nature Switzerland AG 2021
A. van den Essen et al., *Polynomial Automorphisms and the Jacobian Conjecture*,
Frontiers in Mathematics, https://doi.org/10.1007/978-3-030-60535-3

page 114, line 2: $K_{c(s)}$

page 114, line 3: $u := a_2 h^{e-2} + \cdots$

page 114, line 7: So by 5.4.16

page 114, line -6: $c(s+1) + j - c(s) - 1 \leq$

page 124, line -15: of the form $(A(X, Y)^T, Y_1 + M_1(X), \ldots, Y_m + M_m(X))$, where $(X, Y) = (X_1, \ldots, X_n, Y_1, \ldots, Y_m)$, $A \in M_{n \times (n+m)}(R)$

page 164, line -11: degree 4

page 189, line -9: $H = (\ldots, X_4^3, 0)$

page 221, line 44: $= f + D(f)T +$

page 230, line -8: $+(b+1)Y^3 f_2 + \cdots$

page 233, line -10: with $H_n \in B_n$ for all

page 233, line -4: $\ker D_1 \cap B_{n+1}$

page 234, line 5: $= c_{3l+2} \in B_{3l+2}$

page 234, line 6: $= \partial_S^{2l+2} c_{n+1} \in B_l$

page 241, line -12: $h = t(Xt + 1)$

page 248, line 7: integer $s > 1$

page 288, line 2: $f_1 := f - (c_{t_1}/lc(g))sg$. Then either

page 288, line 16: then $G \cap k[Y]$ is a

page 288, line -1: $k[Y] \cap (Y_1 - F_1, \ldots, Y_m - F_m, Q_1, \ldots, Q_s)$

page 289, line -8, -7: $F_1, \ldots, F_m \in k[X] := k[X_1, \ldots, x_n]$ and X_1 algebraic

page 294, line -10: $I_1 = X^2 B$

page 295, line -9: $I_2 = X^3 B$

page 301, line -8: 5.6

page 311, line -18: 551-555

page 313, line -2: Monthly, 70 (1978)

page 314, line -6: fourteenth

page 320, line 3: O. Zariski

page 325, line -19: Kambayashi's

Bibliography

1. S. Abhyankar, P. Eakin, W. Heinzer, On the uniqueness of the coefficient ring in a polynomial ring. J. Algebra **23**, 310–342 (1972)
2. K. Adjamagbo, A. van den Essen, A proof of the equivalence of the Dixmier, Jacobian and Poisson conjectures. Acta Math. Vietnam. **32**(2–3), 205–214 (2007)
3. K. Adjamagbo, A. van den Essen, On the equivalence of the Jacobian, Dixmier and Poisson conjectures in any characteristic. arXiv.org/pdf/math/0608009.pdf
4. T. Asanuma, Polynomial fibre rings of algebras over Noetherian rings. Invent. Math. **87**(1), 101–127 (1987)
5. T. Asanuma, Non-linearizable algebraic group actions on $\mathbf{A}^n$. J. Algebra **166**, 72–79 (1994)
6. T. Bandman, L. Makar-Limanov, Non-stability of the AK invariant. Michigan Math J. **53**, 263–281 (2005)
7. H. Bass, E. Connell, D. Wright, The Jacobian Conjecture: Reduction of Degree and Formal Expansion of the Inverse. Bull. Amer. Math. Soc. (NS) **7**, 287–330 (1982). [MR 83k:14028]
8. A. Belov-Kanel, M. Kontsevich, The Jacobian Conjecture is stably equivalent to the Dixmier Conjecture. Moscow Math. J. Vol. **7**, 209–218 (2007)
9. N. Bourbaki, *Algèbre Commutative* (Hermann, Paris, 1964)
10. D. Cerveau, Dérivations surjectives de l'anneau $\mathbb{C}[x, y]$. J. Algebra **195**, 320–335 (1997)
11. E. Connell, L. van den Dries, Injective polynomial maps and the Jacobian Conjecture. J. Pure Appl. Algebra **28**, 235–239 (1983)
12. D. Cox, J. Little, D. O'Shea, *Using Algebraic Geometry*. Graduate Texts in Mathematics, 2nd edn., vol. 185 (Springer, New York, 2005)
13. A. Crachiola, On automorphisms of Danielewski surfaces. J. Algebraic Geom. **15**(1), 111–132 (2006)
14. A. Crachiola, Cancellation for two-dimensional unique factorization domains. J. Pure Appl. Algebra **213**, 1735–1738 (2009)
15. A. Crachiola, S. Maubach, The Derksen invariant vs. the Makar-Limanov invariant. Proc. Amer. Math. Soc. **131**(11), 3365–3369 (2003)
16. A. Crachiola, L. Makar-Limanov, On the rigidity of small domains. J. Algebra **284**, 1–12 (2005)
17. A. Crachiola, L. Makar-Limanov, An algebraic proof of a cancellation theorem for surfaces. J. Algebra **320**, 3113–3119 (2008)
18. D. Daigle, G. Freudenburg, A counterexample to Hibert's fourteenth problem in dimension 5. J. Algebra **221**, 528–535 (1999)
19. W. Danielewski, On the cancellation problem and automorphism groups of affine algebraic varieties (1989). Preprint, 8 pp.

© The Author(s), under exclusive license to Springer Nature Switzerland AG 2021

A. van den Essen et al., *Polynomial Automorphisms and the Jacobian Conjecture*,
Frontiers in Mathematics, https://doi.org/10.1007/978-3-030-60535-3

20. M. de Bondt, Homogeneous Keller maps. Ph.D. thesis, Radboud University Nijmegen (2009, July). Available at http://webdoc.ubn.ru.nl/mono/b/bondt_m_de/homokema.pdf
21. M. de Bondt, A few remarks on the generalized vanishing conjecture. Arch. Math (Basel) **100**(6), 533–538 (2013)
22. M. de Bondt, Symmetric Jacobians. Cent. Eur. J. Math. **12**(6), 787–800 (2014)
23. M. de Bondt, Mathieu subspaces of codimension less than n of $Mat_n(K)$. Linear Multilinear Algebra **64**(10), 2049–2067 (2016)
24. M. de Bondt, A. van den Essen, Singular Hessians. J. Algebra **282**(1), 195–204 (2004)
25. M. de Bondt, A. van den Essen, A reduction of the Jacobian Conjecture to the symmetric case. Proc. Amer. Math. Soc. **133**(8), 2201–2205 (2005). [MR2138860]
26. M. de Bondt, A. van den Essen, Nilpotent symmetric Jacobian matrices and the Jacobian Conjecture II. J. Pure Appl. Algebra **196**(2–3), 135–148 (2005)
27. H.G.J. Derksen, The kernel of a derivation. J. Pure Appl. Algebra **84**(1), 13–16 (1993)
28. H. Derksen, More on the hypersurface $x + x^2 y + z^2 + t^3 = 0$ in $\mathbf{C}^4$ (1995). Preprint
29. H. Derksen, G. Kemper, Computational invariant theory, in *Invariant Theory and Algebraic Transformation Groups I*. Encyclopaedia of Mathematical Sciences, vol. 130 (Springer, Berlin, 2002)
30. H. Derksen, O. Hadas, L. Makar-Limanov, Newton polytopes of invariants of additive group actions. J. Pure Appl. Algebra **156**, 187–197 (2001)
31. H. Derksen, A. van den Essen, W. Zhao, The Gaussian Moments Conjecture and the Jacobian Conjecture. Israel J. Math. **219**, 917–928 (2017)
32. J. Deveney, D. Finston, G_a-actions on $\mathbb{C}^3$ and $\mathbb{C}^7$. Commun. Algebra **22**(15), 6295–6302 (1994)
33. J. Dixmier, Sur les algèbres de Weyl. Bull. Soc. Math. France **96**, 209–242 (1968)
34. A. Dubouloz, Danielewski-Fieseler surfaces. Transform. Groups **10**(2), 139–162 (2005)
35. A. Dubouloz, Embeddings of Danielewski surfaces in affine spaces. Comment. Math. Helv. **81**(1), 49–73 (2006)
36. A. Dubouloz, Additive group actions on Danielewski varieties and the cancellation problem. Math. Z. **255**(1), 77–93 (2007)
37. A. Dubouloz, K. Palka, The Jacobian Conjecture fails for pseudo-planes. Adv. Math. **339**, 248–284 (2018)
38. A. Dubouloz, P.-M. Poloni, On a class of Danielewski surfaces in affine 3-space. J. Algebra **321**(7), 1797–1812 (2009)
39. E. Dufresne, A. Maurischat, On the finite generation of additive group invariants in positive characteristic. J. Algebra **324**, 1952–1963 (2010)
40. J. Duistermaat, W. van der Kallen, Constant terms in powers of a Laurent polynomial. Indag. Math. (N.S.) **9**(2), 221–231 (1998)
41. C. Dunkl, Y. Xu, *Orthogonal Polynomials in Several Variables*. Encyclopedia of Mathematics and its Applications, vol. 81 (Cambridge University Press, Cambridge, 2001)
42. E. Edo, A. van den Essen, The strong factorial conjecture, J. Algebra **397**, 443–456 (2014)
43. S. Endo, T. Miyata, Invariants of finite abelian groups. J. Math. Soc. Japan **25**, 7–26 (1973)
44. K.-H. Fieseler, On complex affine surfaces with $\mathbb{C}^+$-action. Comment. Math. Helv. **69**, 5–27 (1994)
45. D. Finston, S. Maubach, The automorphism group of certain factorial threefolds and a cancellation problem. Israel J. Math. **163**, 369–381 (2008)
46. J.P. Francoise, F. Pakovich, Y. Yomdin, W. Zhao, Moment vanishing problem and positivity: some examples. Bull. Sci. Math. **135**(1), 10–32 (2011). Also available at: http://www.math.bgu.ac.il/~pakovich/Publications/FPYZ.pdf

47. G. Freudenburg, A counterexample to Hilbert's fourteenth problem in dimension six. Transform. Groups **5**, 61–71 (2000)
48. G. Freudenburg, *Algebraic Theory of Locally Nilpotent Derivations*. Encyclopaedia of Mathematical Sciences, vol. 136 (Springer, Berlin, 2006)
49. G. Freudenburg, L. Moser-Jauslin, Embeddings of Danielewski surfaces. Math. Z. **245**(4), 823–834 (2003)
50. T. Fujita, On Zariski problem. Proc. Japan Acad. Ser. A Math. Sci. **55**, 106–110 (1979)
51. J.-F. Furter, Polynomial composition rigidity and plane polynomial automorphisms (2013). http://perso.univ-lr.fr/jpfurter/
52. J.-F. Furter, H. Kraft, On the geometry of automorphism groups of affine varieties. https://arxiv.org/pdf/1809.04175.pdf
53. C. Godbillon, *Géometrie différentielle et Mécanique* (Hermann, Paris, 1969)
54. N. Gupta, On the cancellation problem for the affine space $\mathbb{A}^3$ in characteristic p. Invent. Math. **195**(1), 279–288 (2014)
55. N. Gupta, On the family of affine threefolds $x^m y = F(x, z, t)$. Compos. Math. **150**(6), 979–998 (2014)
56. N. Gupta, On Zariski's cancellation problem in positive characteristic. Adv. Math. **264**, 296–307 (2014)
57. N. Gupta, A survey on Zariski cancellation problem. Indian J. Pure Appl. Math. **46**(6), 865–877 (2015)
58. R. Gurjar, A topological proof of cancellation theorem for $\mathbf{C}^2$. Math. Z. **240**, 83–94 (2002)
59. O. Hadas, On the vertices of Newton polytopes associated with an automorphism of the ring of polynomials. J. Pure Appl. Algebra **76**, 81–86 (1991)
60. O. Hadas, L. Makar-Limanov, Newton polytopes of constants of locally nilpotent derivations. Commun. Algebra **28**(8), 3667–3678 (2000)
61. H. Hasse, F.K. Schmidt, Noch eine Begründung der Theorie der höheren Differentialquotienten in einem algebraischen Funktionenkörper einer Unbestimmten. J. Reine Angew. Math. **177**, 215–237 (1937)
62. S. Iitaka, T. Fujita, Cancellation theorem for algebraic varieties. J. Fac. Sci. Univ. Tokyo Sect. IA Math. **24**, 123–127 (1977)
63. H. Jung, Über ganze birationale Transformationen der Ebene. J. Reine Angew. Math. **184**, 161–174 (1942)
64. H. Kojima, M. Miyanishi, On Roberts' counterexample to the fourteenth problem of Hilbert. J. Pure Appl. Algebra **122**, 277–292 (1997)
65. A. Konijnenberg, Mathieu subspaces of finite products of matrix rings. Master's thesis, Radboud University Nijmegen, August (2012)
66. K. Kurano, Positive characteristic finite generation of symbolic Rees algebras and Roberts' counterexamples to the fourteenth problem of Hilbert. Tokyo J. Math. **16**(2), 473–496 (1993)
67. S. Kuroda, The infiniteness of the SAGBI bases for certain invariant rings. Osaka J. Math. **39**(3), 665–680 (2002)
68. S. Kuroda, A generalization of Roberts' counterexample to the fourteenth problem of Hilbert. Tohoku Math. J. **56**, 501–522 (2004)
69. S. Kuroda, A counterexample to the fourteenth problem of Hilbert in dimension four. J. Algebra **279**(1), 126–134 (2004)
70. S. Kuroda, A counterexample to the fourteenth problem of Hilbert in dimension three. Michigan Math. J. **53**(1), 123–132 (2005)
71. S. Kuroda, Fields defined by locally nilpotent derivations and monomials. J. Algebra **293**(2), 395–406 (2005)

72. S. Kuroda, Hilbert's fourteenth problem and algebraic extensions. J. Algebra **309**(1), 282–291 (2007)
73. S. Kuroda, A generalization of the Shestakov-Umirbaev inequality. J. Math. Soc. Japan **60**, 495–510 (2008)
74. S. Kuroda, Shestakov-Umirbaev reductions and Nagata's conjecture on a polynomial automorphism. Tohoku Math. J. **62**, 75–115 (2010)
75. S. Kuroda, An algorithm for deciding tameness of polynomial automorphisms in three variables, in *Commutative Algebra and Algebraic Geometry (CAAG-2010)*. Ramanujan Mathematical Society Lecture Notes Series, vol. 17 (Ramanujan Mathematical Society, Mysore, 2010), pp. 117–137
76. S. Kuroda, Hilbert's fourteenth problem and field modifications. J. Algebra **556**, 93–105 (2020)
77. L. Makar-Limanov, On the hypersurface $x + x^2y + z^2 + t^3 = 0$ in $\mathbf{C}^4$, or a $\mathbf{C}^3$-like threefold which is not $\mathbf{C}^3$. Israel J. Math. **96**, 419–429 (1996)
78. L. Makar-Limanov, Locally nilpotent derivations, a new ring invariant and applications. Lecture notes (1998). Available at http://www.math.wayne.edu/~lml
79. L. Makar-Limanov, On the group of automorphisms of a surface $x^n y = P(z)$. Israel J. Math. **121**, 113–123 (2001)
80. L. Makar-Limanov, Locally nilpotent derivations on the surface $xy = p(z)$, in *Proceedings of the Third International Algebra Conference (Tainan, 2002)* (Kluwer Academic Publisher, Dordrecht, 2003), pp. 215–219
81. L. Makar-Limanov, *Again $x + x^2y + z^2 + t^3 = 0$*. Contemporary in Mathematics, vol. 369 (American Mathematical Society, Providence, 2005)
82. O. Mathieu, *Some Conjectures about Invariant Theory and Their Applications*. Algèbre non commutative, groupes quantiques et invariants (Reims, 1995), Sémin. Congr. 2, Soc. Math. (1997), pp. 263–279
83. H. Matsumura, *Commutative ring Theory*. Cambridge Studies in Advanced Mathematics, vol. 8 (Cambridge University Press, Cambridge, 1989)
84. G. Meng, Legendre transform, Hessian conjecture and tree formula. Appl. Math. Lett. **19**(6), 503–510 (2006)
85. M. Miyanishi, G_a-action of the affine plane. Nagoya Math. J. **41**, 97–100 (1971)
86. M. Miyanishi, *Curves on Rational and Unirational Surfaces*. Tata Institute of Fundamental Research Lectures on Mathematics and Physics, vol. 60 (Tata Institute of Fundamental Research, Bombay, 1978)
87. M. Miyanishi, Lectures on geometry and topology of polynomials surrounding the Jacobian conjecture (2015). https://arxiv.org/pdf/1504.07179.pdf
88. M. Miyanishi, T. Sugie, Affine surfaces containing cylinderlike open sets. J. Math. Kyoto Univ. **20**, 11–42 (1980)
89. L. Moser-Jauslin, P.-M. Poloni, Embeddings of a family of Danielewski hypersurfaces and certain $\mathbf{C}^+$-actions on $\mathbf{C}^3$. Ann. Inst. Fourier (Grenoble) **56**(5), 1567–1581 (2006)
90. S. Mukai, *Counterexample to Hilbert's Fourteenth Problem for the 3-dimensional additive Group*. RIMS Preprint 1343 (Kyoto Univ., Res. Inst. Math. Sci., Kyoto, 2001)
91. M. Nagata, On the 14-th problem of Hilbert. Amer. J. Math. **81**, 766–772 (1959)
92. M. Nagata, On the fourteenth problem of Hilbert, in *Proceedings of the International Congress of Mathematicians, 1958* (Cambridge University Press, London, 1960), pp. 459–462
93. M. Nagata, *Lectures on the Fourteenth Problem of Hilbert* (Tata Institute of Fundamental Research, Bombay, 1965)
94. M. Nagata, *On Automorphism Group of $k[x, y]$*. Kyoto University Lectures in Mathematics, vol. 5 (Kyoto University, Kinokuniya, Tokyo, 1972)

95. S. Nieman, Mathieu subspaces of univariate polynomial rings. Master's Thesis, Radboud University Nijmegen, July (2012)

96. E. Noether, Der Endlichkeitssatz der Invarianten endlicher Gruppen. Math. Ann. **77**(1), 89–92 (1915)

97. E. Noether, Gleichungen mit vorgeschriebener Gruppe. Math. Ann. **78**(1), 221–229 (1917)

98. A. Nowicki, *Polynomial Derivations and their Rings of Constants* (Uniwersytet Mikołaja Kopernika, Toruń, 1994)

99. R. Peretz, *The Asymptotic Variety of Polynomial Maps* (Lambert Academic Publishers, Saarbrücken, 2016)

100. P.-M. Poloni, Sur les plongements des hypersurfaces de Danielewski. Ph.D. thesis, Université de Bourgogne (2008)

101. T. Reijnders, The generalized vanishing conjecture and the factorial conjecture. Master's Thesis, Radboud University Nijmegen, August (2012)

102. R. Rentschler, Opérations du groupe additif sur le plan affine. C.R. Acad. Sci. Paris **267**, 384–387 (1968)

103. L. Robbiano, M. Sweedler, *Subalgebra bases, in Commutative Algebra*, ed. by W. Bruns, A. Simis. Lecture Notes in Mathematics, vol. 1430 (Springer, Berlin, 1988), pp. 61–87

104. P. Roberts, An infinitely generated symbolic blow-up in a power series ring and a new counterexample to Hibert's fourteenth problem. J. Algebra **132**, 461–473 (1990)

105. B. Rocks, Incompatibility of Diophantine equations arising from the strong factorial conjecture. Ph. D. thesis, May 2015, Washington University in St. Louis (2015)

106. P. Russell, On affine-ruled rational surfaces. Math. Ann. **255**, 287–302 (1981)

107. P. Russell, *Open Problems in Affine Algebraic Geometry* (collected by G. Freudenburg and P. Russell). Contemporary in Mathematics, vol. 369 (American Mathematical Society, Providence, 2005)

108. A. Sathaye, Polynomial ring in two variables over a DVR: a criterion. Invent. Math. **74**(1), 159–168 (1983)

109. I. Shestakov, U. Umirbaev, Poisson brackets and two-generated subalgebras of rings of polynomials. J. Amer. Math. Soc. **17**, 181–196 (2004)

110. I. Shestakov, U. Umirbaev, The tame and the wild automorphisms of polynomial rings in three variables. J. Amer. Math. Soc. **17**, 197–227 (2004)

111. Y. Stein, On the density of image of differential operators generated by polynomials. J. Anal. Math. **52**, 291–300 (1989)

112. R. Steinberg, Nagata's example, in *Algebraic Groups and Lie Groups*. Australian Mathematical Society Lecture Series, vol. 9 (Cambridge University Press, Cambridge, 1997), pp. 375–384

113. R.G. Swan, Invariant rational functions and a problem of Steenrod. Invent. Math. **7**, 148–158 (1969)

114. G. Szegö, *Orthogonal Polynomials*, 4th edn., vol. XXIII (American Mathematical Society, Colloquium Publications, Providence, 1975)

115. B. Totaro, Hilbert's 14th problem over finite fields and a conjecture on the cone of curves. Compos. Math. **144**(5), 1176–1198 (2008)

116. Y. Tsuchimoto, Endomorphisms of the Weyl algebra and p-curvatures. Osaka J. Math. **42**, 435–452 (2005)

117. A. van den Essen, *Polynomial Automorphisms and the Jacobian Conjecture*, Progress in Mathematics, vol. 190 (Birkhäuser Verlag, Basel, 2000)

118. A. van den Essen, The amazing image conjecture (2010). Preprint. arXiv:1006.5801v1

119. A. van den Essen, *An Introduction to Mathieu Subspaces*. Lectures delivered at the Chern Institute of Mathematics, Tianjin (2014, July). arXiv:1907.06107v1

120. A. van den Essen, R. Lipton, A p-adic approach to the Jacobian Conjecture. J. Pure Appl. Algebra **219**(7), 2624–2628 (2015)
121. A. van den Essen, S. Nieman, Mathieu-Zhao spaces of univariate polynomial rings with non-zero strong radical. J. Pure Appl. Algebra **220**, 3300–3306 (2016)
122. A. van den Essen, L. van Hove, Mathieu-Zhao spaces of polynomial rings (2019). arXiv:1907.06106v1
123. A. van den Essen, S. Washburn, The Jacobian Conjecture for symmetric matrices. J. Pure Appl. Algebra **189**(1–3), 123–133 (2004)
124. A. van den Essen, W. Zhao, Two results on homogeneous Hessian nilpotent polynomials. J. Pure Appl. Algebra **212**(10), 2190–2193 (2008). See also arXiv: 0704.1690v1
125. A. van den Essen, W. Zhao, Mathieu subspaces of univariate polynomial algebras. J. Pure Appl. Algebra **217**(7), 1316–1324 (2013)
126. A. van den Essen, D. Wright, W. Zhao, Images of locally finite derivations of polynomial algebras in two variables. J. Pure Appl. Algebra **215**, 2130–2134 (2011). See also arXiv:1004.0521v1 [math.AC]
127. A. van den Essen, D. Wright, W. Zhao, On the image conjecture. J. Algorithms **340**(1), 211–224 (2011). See also arXiv:1008.3962 [math.RA]
128. A. van den Essen, R. Willems, W. Zhao, Some results on the vanishing conjecture of differential operators with constant coefficients. J. Pure Appl. Algebra **219**(9), 3847–3861 (2015)
129. W. van der Kulk, On polynomial rings in two variables. Nieuw Arch. Wisk. (3) **1**, 33–41 (1953)
130. L. van Hove, Mathieu-Zhao subspaces. Master's thesis Radboud University Nijmegen, July (2015)
131. B. Veĭsfeĭler, I. Dolgachev, Unipotent group schemes over integral rings (Russian). Izv. Akad. Nauk SSSR Ser. Mat. **38**, 757–799 (1974)
132. J. Wilkens, On the cancellation problem for surfaces. C. R. Acad. Sci. Paris Sér. I Math. **326**(9), 1111–1116 (1998)
133. A. Zaks, Dedekind subrings of $k[x_1, \cdots, x_n]$ are rings of polynomials. Israel J. Math. **9**, 285–289 (1971)
134. O. Zariski, Interprétations algébrico-géométriques du quatorzième problème de Hilbert. Bull. Sci. Math. **78**, 155–168 (1954)
135. W. Zhao, Hessian nilpotent polynomials and the Jacobian Conjecture. Trans. Amer. Math. Soc. **359**(1), 294–274 (2007). See also math.CV/0409534
136. W. Zhao, A vanishing conjecture on differential operators with constant coefficients. Acta Math. Vietnam **32**(3), 259–286 (2007). See also arXiv:0704.1691v2
137. W. Zhao, Images of commuting differential operators of order one with constant leading coefficients. J. Algebra **324**(2), 231–247 (2010). See also arXiv:0902.0210 [math.CV]
138. W. Zhao, Generalizations of the image conjecture and the Mathieu conjecture. J. Pure Appl. Algebra **214**, 1200–1216 (2010). See also arXiv:0902.0212 [math.CV]
139. W. Zhao, Mathieu subspaces of associative algebras. J. Algebra **350**(1), 245–272 (2012). See also arXiv: 1005.4260
140. W. Zhao, R. Willems, An analogue of the Duistermaat-van der Kallen theorem for group algebras. Central Eur. J. Math. **10**(3), 974–986 (2012)

Index